PREFACE

These notes arose from a one-semester course in the foundations of projective geometry, given at Harvard in the fall term of 1966-1967.

We have approached the subject simultaneously from two different directions. In the purely synthetic treatment, we start from axioms, and build the abstract theory from there. For example, we have included the synthetic proof of the fundamental theorem for projectivities on a line, using Pappus' axiom. On the other hand we have the real projective plane as a model, and use methods of Euclidean geometry or analytic geometry to see what is true in that case. These two approaches are carried along independently, until the first is specialized by the introduction of more axioms, and the second is generalized by working over an arbitrary field or division ring, to the point where they coincide in chapter 7, with the introduction of coordinates in an abstract projective plane.

Throughout the course there is special emphasis on the various groups of transformations which arise in projective geometry. Thus the reader is introduced to group theory in a practical context. We do not assume any previous knowledge of algebra, but do recommend a reading assignment in abstract group theory, such as [4].

There is a small list of problems at the end of the notes , which should be taken in regular doses along with the text .

There is also a small bibliography , mentioning various works referred to in the preparation of these notes. However, I am most indebted to Oscar Zariski , who taught me the same course eleven years ago.

R. Hartshorne
March 1967.

CONTENTS

CHAPTER 1.

INTRODUCTION : AFFINE PLANES AND PROJECTIVE PLANES.

Projective geometry is concerned with properties of incidence. Properties which are invariant under stretching, translation, or rotation of the plane. Thus in the axiomatic development of the theory, the notions of distance and angle will play no part.

However, one of the most important examples of the theory is the real projective plane, and there we will use all the techniques available (e.g. those of Euclidean geometry and analytic geometry) to see what is true and what is not true.

Affine geometry

Let us start with some of the most elementary facts of ordinary plane geometry, which we will take as axioms for our synthetic development.

Definition. An affine plane is a set, whose elements are called points, and a set of subsets, called lines, satisfying the following three axioms, A1 - A3. We will use the terminology " P lies on l ", or " l passes through P " to mean the point P is an element of the line l.

A 1. Given two distinct points P and Q, there is one and only one line containing both P and Q.

We say that two lines are parallel if they are equal, or if they have no points in common.

A 2. Given a line l and a point P, not on l, there is one and only one line m, which is parallel to l, and which passes through P.

A 3. There exist three non-collinear points. (A set of points $P_1, \ldots P_n$ is said to be collinear if there exists a line l containing them all.)

Notation.

$P \neq Q$	P is not equal to Q.
$P \in l$	P lies on l.
$l \cap m$	the intersection of l and m.
$l \parallel m$	l is parallel to m.
$\forall$	for all.
$\exists$	there exists.
$\Rightarrow$	implies.
$\Leftrightarrow$	if and only if.

Example : The ordinary plane, known to us from Euclidean geometry, satifies the axioms A 1 - A 3, and therefore is an affine plane.

A convenient way of representing this plane is by introducing Cartesian coordinates, as in analytic geometry. Thus a point P is represented as a pair (x, y) of real numbers. (We write $x, y \in \mathbb{R}$.)

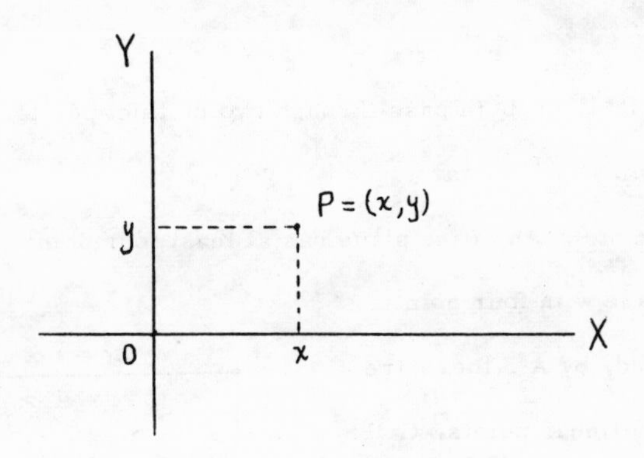

Proposition 1.1 Parallelism is an equivalence relation.

Definition. A relation $\sim$ is an equivalence relation if it has the following three properties :

1. Reflexive : $a \sim a$
2. Symmetric : $a \sim b \Rightarrow b \sim a$
3. Transitive : $a \sim b$ and $b \sim c \Rightarrow a \sim c$.

Proof of Proposition : We must check the three properties

1. Any line is parallel to itself, by definition.
2. $l \parallel m \Rightarrow m \parallel l$ by definition

3. If $l \parallel m$, and $m \parallel n$, we wish to prove $l \parallel n$. If $l = n$, there is nothing to prove. If $l \neq n$, and there is a point $P \in l \cap n$, then l, n are both $\parallel m$, and pass through P, which is impossible, by axiom A 2. We conclude that $l \cap n = \emptyset$ (the empty set), and so $l \parallel n$.

Proposition 1.2 Two distinct lines have at most one point in common.

For if l, m both pass through two distinct points P, Q, then by axiom A 1, $l = m$.

Example : An affine plane has at least four points. There is an affine plane with four points.

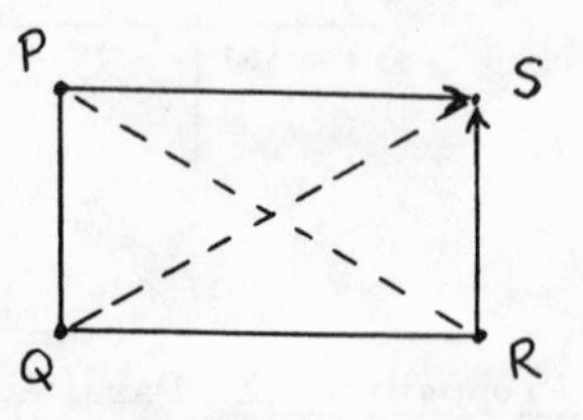

Indeed, by A 3 there are three non-collinear points. Call them P, Q, R. By A 2 there is a line l through P, parallel to the line QR joining Q, and R, which exists by A 1. Similarly, there is a line $m \parallel PQ$, passing through R.

Now l is not parallel to m ($l \nparallel m$). For if it were, then we would have

$$PQ \parallel m \parallel l \parallel QR$$

and hence $PQ \parallel QR$ by Proposition 1.1. This is impossible, however, because $PQ \neq QR$, and both contain Q.

Hence l must meet m in some point S. Since S lies on m, which is parallel to PQ, and different from PQ, S does not lie on PQ, so $S \neq P$, and $S \neq Q$. Similarly $S \neq R$. Thus S is indeed a fourth point. This proves the first assertion.

Now consider the lines PR and QS. It may happen that they meet (for example in the real projective plane they will (proof ?)). On the other hand, it is consistent with the axioms to assume that they do not meet.

In that case we have an affine plane consisting of four points P, Q, R, S, and six lines PQ, PR, PS, QR, QS, RS, and one can verify easily that the axioms A 1 - A 3 are verified. This is the smallest affine plane.

Definition. A pencil of lines is either a) the set of all lines passing through some point P, or b) the set of all lines parallel to some line l. In the second case we speak of a pencil of parallel lines.

Definition. A one-to-one correspondence between two sets X and Y is a mapping $T : X \longrightarrow Y$ (i.e. a rule T, which associates to each element x of the set X an element $T(x) = y \in Y$) such that $x_1 \neq x_2 \Rightarrow Tx_1 \neq Tx_2$, and $\forall y \in Y, \exists x \in X$ such that $T(x) = y$.

Ideal points and the projective plane.

We will now complete the affine plane by adding certain "points at infinity" and thus arrive at the notion of the projective plane.

Let A be an affine plane. For each line $l \in A$, we will denote by $[l]$ the pencil of lines parallel to l, and we will call $[l]$ and ideal point, or point at infinity, in the direction of l. We write $P^* = [l]$.

We define the completion S of A as follows. The points of S are the points of A, plus all the ideal points of A. A line in S is either

a) An ordinary line l of A, plus the ideal point $P^* = [l]$ of l, or

b) the "line at infinity", consisting of all the ideal points of A.

We will see shortly that S is a projective plane, in the sense of the following definition.

Definition A projective plane S is a set, whose elements are called of points, and a set of subsets, called lines, satisfying the following four axioms.

P 1. Two distinct points P, Q of S lie on one and only one line.

P 2. Any two lines meet in at least one point.

P 3. There exist three non-collinear points.

P 4. Every line contains at least three points.

<u>Proposition 1.3</u> The completion S of an affine plane A, as described above, is a projective plane.

<u>Proof:</u> We must verify the four axioms P 1 - P 4 of the definition.

<u>P 1.</u> Let $P, Q \in S$. 1) if P, Q are ordinary points of A, then P and Q lie on only one line of A. They do not lie on the line at infinity of S, hence they lie on only one line of S.

2) if P is an ordinary point, and $Q = [l]$ is an ideal point, we can find by A 2 a line m such that $P \in m$, and $m \parallel l$, i.e. $m \in [l]$, so that Q lies on the extionsion of m to S. This is clearly the only line of S containing P and Q.

3) if P, Q are both ideal points, then they both lie on the line of S containing them.

<u>P 2.</u> Let l, m be lines. 1) If they are both ordinary lines, and $l \not\parallel m$, then they meet in a point of A. If $l \parallel m$, then the ideal point $P^* = [l] = [m]$ lies on both l and m in S.

2) if l is an ordinary line, and m is the line at infinity, then $P^* = [l]$ lies on both l and m.

<u>P 3.</u> Follows immediately from A3. One must check only that if P, Q, R are non-collinear in A, then they are also non-collinear in S. Indeed, the only new line is the line at infinity, which contains none of them.

P 4. Indeed, by problem number 1, it follows that each line of A contains at least two points. Hence, in S it has also its point at infinity, so has at least three points.

Examples : 1. By completing the real affine plane of Euclidean geometry, we obtain the real projective plane.

2. By completing the affine plane of 4 points, we obtain a projective plane with 7 points.

3. Another example of a projective plane can be constructed as follows: let $\mathbb{R}^3$ be ordinary Euclidean 3-space, and let O be a point of $\mathbb{R}^3$. Let L be the set of lines through O.

We define a point of L to be a line through O in $\mathbb{R}^3$. We define a line of L to be the collection of lines through O which all lie in some plane through O.

Then L satisfies the axioms P 1 - P 4, (left to reader), and so it is a projective plane.

Homogeneous coordinates in the real projective plane.

We can give an analytic definition of the real projective plane as follows. We consider the example given above of lines in $\mathbb{R}^3$. A point of S is a line through O. We will represent the point P of S corresponding to l by choosing any point (x_1, x_2, x_3) on l different from the point (o, o, o). The numbers x_1, x_2, x_3 are homogeneous coordinates of P. Any other point of l has the coordinates $(\lambda x_1, \lambda x_2, \lambda x_3)$, where $\lambda \in \mathbb{R}$, $\lambda \neq 0$. Thus S is the collection of triples (x_1, x_2, x_3) of real numbers, not all zero, and two triples (x_1, x_2, x_3) and (x_1', x_2', x_3') represent the same point $\Leftrightarrow \exists\, \lambda \in \mathbb{R}$ such that

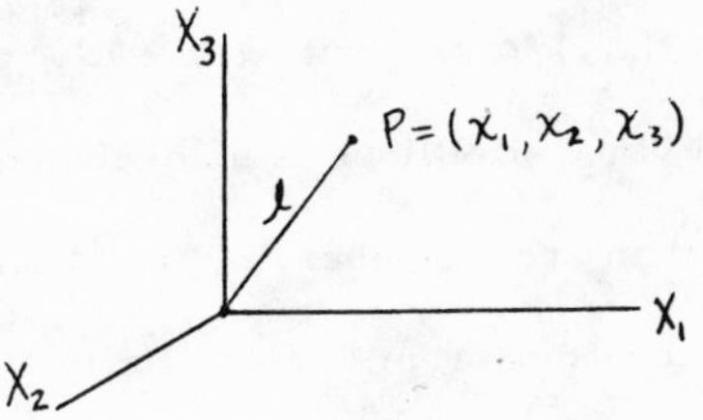

$$x_i' = \lambda x_i \qquad \text{for } i = 1, 2, 3.$$

Since the equation of a plane in $\mathbb{R}^3$ passing through O is of the form

$$a_1 x_1 + a_2 x_2 + a_3 x_3 = 0 \qquad a_i \text{ not all } 0,$$

we see that this is also the equations of a line of S, in terms of the homogeneous coordinates.

Definition. Two projective planes S and S' are isomorphic if there exists a one-to-one transformation $T : S \longrightarrow S'$ which takes collinear points into collinear points.

Proposition 1.4 The projective plane S defined by homogeneous coordinates which are real numbers, as above, is isomorphic to the projective plane obtained by completing the ordinary affine plane of Euclidean geometry.

Proof: On the one hand, we have S, whose points are given by homogeneous coordinates $(x_1, x_2, x_3,)$, $x_i \in \mathbb{R}$, not all zero. On the other hand, we have the Euclidean plane A, with Cartesian coordinates x, y. Let us call its completion S'. Thus the points of S' are the points (x, y) of A (with $x, y \in \mathbb{R}$), plus the ideal points. Now a pencil of parallel lines is uniquely determined by its slope m, which may be any real number, or ∞. Thus the ideal points are described by the coordinate m.

Now we will define a mapping $T : S \longrightarrow S'$ which will exhibit the isomorphism of S and S'. Let $(x_1, x_2, x_3) = P$ be a point of S.

1) If $x_3 \neq o$, we define $T(P)$ to be the point of A with coordinates $x = x_1/x_3$, $y = x_2/x_3$. Note that this is uniquely determined, because if we replace (x_1, x_2, x_3) by $(\lambda x_1, \lambda x_2, \lambda x_3)$, then x and y do not change. Note also that every point of A can be obtained in this way. Indeed, the point with coordinates (x, y) is the image of the point of S with homogeneous coordinates $(x, y, 1)$.

2) If $x_3 = o$, then we define $T(P)$ to be the ideal point of S' with slope $m = x_2/x_1$. Note that this makes sense, because x_1 and x_2 cannot both be zero. Again replacing (x_1, x_2, o) by $(\lambda x_1, \lambda x_2, o)$ does not change m. Also each value of m occurs : If $m \neq \infty$, we take $T(1, m, o)$, and if $m = \infty$, we take $T(o, 1, o)$.

Thus T is a one-to-one mapping of S into S'. We must check that T takes collinear points into collinear points. A line l in S is given by an equation

$$a_1x_1 + a_2x_2 + a_3x_3 = o$$

1) Suppose that a_1 and a_2 are not both zero. Then for those points with $x_3 = o$, namely the point given by $x_1 = \lambda a_2$, $x_2 = -\lambda a_1$,

T of this point is the ideal point given by the slope $m = -a_1/a_2$, which indeed is on a line in S' with the finite points.

2) If $a_1 = a_2 = o$, l has the equation $x_3 = o$. Any point of S with $x_3 = o$ goes to an ideal point of S', and these form a line.

q. e. d.

Remark : From now on, we will not distinguish between the two isomorphic planes of Proposition 1.4, and will call them (or it) the real projective plane. It will be the most important example of the axiomatic theory we are going to develop, and we will often check results of the axiomatic theory in this plane by way of example. Similarly, theorems in the real projective plane can geve motivation for results in the axiomatic theory. However, to establish a theorem in our theory, we must derive it from the axioms and from previous theorems. If we find that it is true in the real projective plane, that is evidence in favor of the theorem, but does not constitute a proof in our set-up.

Note also that if we remove any line from the real projective plane, we obtain the Euclidean plane.

CHAPTER 2 DESARGUES THEOREM.

One of the first main results of projective geometry is "Desargues theorem", which states the following :

P 5. Desargues theorem :

Let two triangles ABC, and A'B'C' be such that the lines joining corresponding vertices, namely AA', BB' and CC', pass through a point O. (We say that the two triangles are perspective from O.) Then the three pairs of corresponding sides intersect in three points

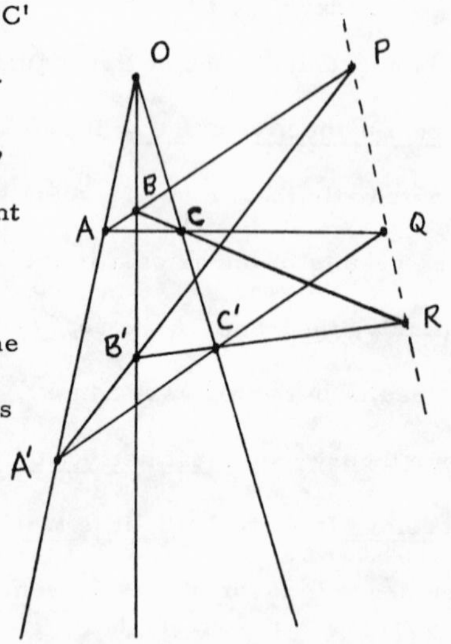

P = AB. A'B'

R = BC. B'C'

Q = AC. A'C',

which lie in a straight line.

Now it is not quite right for us to call this a "theorem", because it cannot be proved form our axioms P 1 - P 4. However, we will show that it is true in the real projective plane (and more generally, for any projective plane which can be embedded in a three-dimensional projective space.) Then we will take this statement as a further axiom, P 5, of our abstract projective geometry. We will show by an example that P 5 is not a consequence of P 1 - P 4 : namely, we will exhibit a geometry which satisfies P1 - P 4; but not P 5.

Definition : A projective 3-space is a set whose elements are called points, together with certain subsets called lines, and certain other subsets called planes, which satisfies the following axioms :

S 1. Two distinct points P, Q lie on one and only one line l.

S 2. Three non-collinear points P, Q, R, lie on a unique plane.

S 3. A line meets a plane in at least one point.

S 4. Two planes have at least a line in common.

S 5. There exist four non-coplanar points , no three of which are collinear.

S 6. Every line has at least three points.

Example : By a process analogous to that of completing an affine plane to a projective plane, the ordinary Euclidean three-space can be completed to a projective three-space , which we call real projective three-space . Alternatively , this same real projective three-space can be described by homogeneous coordinates, as follows. A point is described by a quadruple (x_1, x_2, x_3, x_4) of real numbers, not all zero, where we agree that (x_1, x_2, x_3, x_4) and $(\lambda x_1, \lambda x_2, \lambda x_3, \lambda x_4)$ re present the same point, for any $\lambda \in \mathbb{R}$, $\lambda \neq 0$. A plane is defined by a linear equation

$$\sum_{i=1}^{4} a_i x_i = 0 \qquad\qquad a_i \in \mathbb{R} ,$$

and a line is defined as the intersection of two distinct planes. The details of verification of the axioms are left to the reader.

Now the remarkable fact is that although P 5 is not a consequence of P 1 - P 4 in the projective plane, it is a consequence of the seemingly equally simple axioms for projective three-space.

Theorem 2.1 Desargues theorem is true in projective three-space, where we do not necessarily assume that all the points lie in a plane. In particular, Desargues theorem is true for any plane (which by problem number 8 is a projective plane) which lies in a projective three-space.

Proof: Case 1. : Let us assume that the plane Σ containing the points A, B, C is different form the plane Σ' containing the points A', B', C'. The lines AB and A'B' both lie in the plane determined by O, A, B, and so they meet in a point P. Similarly we see that AC and A'C' meet, and that BC and B'C' meet. Now the points P, Q, R lie in the plane Σ, and also in the plane Σ'. Hence they lie in the intersection $\Sigma \cap \Sigma$, which is a line (Problem 7c)

Case 2. Suppose that $\Sigma = \Sigma'$, so that the whole configuration lies in one plane (call it Σ) . Pick a point X which does not lie in Σ (this exists by axiom S 5) . Draw lines joining X to all the points in the diagram. Choose D on XB , different from B, and let D' = OD.XB'. (Why do they meet ?) Then the triangles ADC and A'D'C' are perspective from O, and do not lie in the same plane. We conclude from Case 1 That the points

$$P' = AD.A'D'$$

$$Q = AC.A'C'$$

$$R' = DC.D'C'$$

lie in a line. But these points are projected for X into P, Q, R, on Σ, hence P, Q, R are collinear.

Remark : The configuration of Desargues theorem has a lot of symmetry. It consists of 10 points and 10 lines. Each point lies on three lines, and each line contains 3 points. Thus it may be given the symbol (10_3). Also , the role of the various points is not fixed. Any one of the ten points can be taken as the center of perspectivity of two triangles. In fact, the group of automorphisms of the configuration is Σ_5 , the symmetric group on 5 letters. (Consider the action of any automorphism on the space version of the configuration. It must permute the five planes OAB, OBC, OAC, ABC A'B'C'.) See problems 12, 13, 14 for details.

We will now give an example of a non-Desarguesian projective plane, that is, a plane satisfying P1, P2, P3, P4, but not P5. This will show that P5 is not a logical consequence of P 1 - P 4.

Definition : A configuration is a set, whose elements are called "points", and a collection of subsets, calles "lines", which satisfies the following axiom :

C 1. Two distinct points lie on at most one line.

It follows that two distinct lines have at most one point in common.

Examples : Any affine plane or projective plane is a configuration Any set of "points" , and no lines is a configuration. The collection of 10 points and 10 lines which occurs in Desargues theorem is a configuration.

Let π_o be a configuration. We will now define the free projective plane generated by π_o .

Let π_1 be the new configuration defined as follows : The points of π_1 are the points of π_o . The lines of π_1 are the lines of π_o , plus, for each pair of points $P_1, P_2 \in \pi_o$ not on a line, a new line $\{P_1, P_2\}$. Then π_1 has the property

a) Every two distinct points lie on a line.

Construct π_2 from π_1 as follows. The points of π_2 are the points of π_1 , plus , for each pair of lines l_1, l_2, of π_1 which do not meet, a new point $l_1 \cdot l_2$. The lines of π_2 are the lines of π_1, extended by their new points. e.g. the point $l_1 \cdot l_2$ lies on the extensions of the lines l_1, l_2 . Then π_2 has the property

b) every pair of distinct lines meets in a point,

but π_2 no longer has the property a).

We proceed in the same fashion. For n even, we construct π_{n+1} by adding new lines, and for n odd, we construct π_{n+1} by adding new points.

Let $\Pi = \bigcup_{n=o}^{\infty} \pi_n$, and define a line Π to be a subset of $L \subseteq \Pi$ such that for all large enough n, $L \cap \pi_n$ is a line of π_n .

Proposition 2.2 If π_o contains at least four points, no three of which lie on a line, then Π is a projective plane.

Proof : Note that for n even, π_n satisfies b), and for n odd π_n satisfies a). Hence Π satisfies both a) and b), i.e. it satisfies P 1 and P2. If P, Q, R are three non-collinear points of π_o, then they are also non-collinear in Π. Thus P 3 is satisfied. Axiom P 4 is left to the reader: show each line of π has at least three points.

Definition. A confined configuration is a configuration in which each point is on at least three lines, and each line contains at least three points.

Example : The configuration of Desargues theorem is confined.

Proposition 2.3 Any finite, confined configuration of Π is already contained in π_o.

Proof: For a point $P \in \Pi$ we define its level as the smallest $n \geq o$ such that $P \in \pi_n$. For a line $L \subseteq \Pi$, we define its level to be the smallest $n \geq o$ such that $L \cap \pi_n$ is a line.

Now let Σ be a finite confined configuration in Π, and let n be the maximum level of a point of line in Σ. Suppose it is a line $l \subseteq \Sigma$ which has level n. (A similar argument holds if a point has maximum level.) Then $l \cap \pi_n$ is a line, and $l \cap \pi_{n-1}$ is not a line. If $n = o$, we are done, $\Sigma \subseteq \pi_o$. Suppose $n > o$. Then l occurs as the line joining two points of π_{n-1} which did not lie on a line, But all points of Σ have level $\leq n$, so they are in π_n, so l can contain at most two of them, which is a contradiction.

Example 2.4 A non-Desarguesian projective plane. Let π_o be four points, and no lines. Let Π be the free projective plane generated by π_o. Note as a Corollary of the previous proposition, that Π is infinite, and so every line contains infinitely many points. Thus it is possible to choose O, A, B, C , no three collinear, A' on OA, B' on OB, C' on OC, such that they form 7 distinct points, and A', B', C' are not collinear. Then construct

$$P = AB.A'B'$$

$$Q = AC.A'C'$$

$$R = BC.B'C'$$

Check that all 10 points are distinct. If Desargues theorem is true in Π, then P, Q, R lie on a line, hence these 10 points and 10 lines form a confined configuration, which must lie in π_o, since π_o has only four points.

CHAPTER 3. DIGRESSION ON GROUPS AND AUTOMORPHISMS

Definition : A group is a set G, together with a binary operation, called multiplication, written ab, such that

G 1. (Associativity) For all $a, b, c \in G$,

$$(ab)c = a(bc)$$

G 2. There exists an element $1 \in G$ such that

$$a \cdot 1 = 1 \cdot a = a \qquad \text{for all } a.$$

G 3. For each $a \in G$, there exists an element $a^{-1} \in G$ such that

$$aa^{-1} = a^{-1}a = 1.$$

The element 1 is called the identity, or unit element. The element a^{-1} is called inverse of a.

Note that in general the product ab may be different from ba. However, we say the group G is abelian, or commutative if

G 4. For all $a, b, \in G$, $ab = ba$.

Examples. 1. Let S be any set, and let G be the set of permutations of the set S. A permutation is a 1 - 1 mapping of S onto S. If $g_1, g_2 \in G$ are two permutations, we define $g_1 g_2 \in G$ to be the permutation obtained by performing first g_2, then g_1. (i.e. if $x \in S$,

$$(g_1 g_2)(x) = g_1(g_2(x)). \quad)$$

2. Let C be a configuration, and let G be the set of automorphisms of C, i.e. the set of those permutations of C which

send lines onto lines. Again we define the product $g_1 g_2$ of two automorphisms g_1, g_2, by performing g_2 first, and then g_1. This group is written Aut C.

Definition : A homomorphism $\varphi : G_1 \longrightarrow G_2$ of one group to another, is a mapping of the set G_1 to the set G_2 such that

$$\varphi(ab) = \varphi(a)\varphi(b)$$

for each $a, b \in G_1$.

An isomorphism of one group with another, is a homomorphism which is 1 - 1 and onto.

Definition . Let G be a group. A subgroup of G is a non-empty subset $H \subseteq G$, such that for any $a, b \in H$, $ab \in H$, and $a^{-1} \in H$.

Note this condition implies $1 \in H$.

Example. Let G = Perm S, the group of permutations of a set S, let $x \in S$, and let $H = \{ g \in G \mid g(x) = x \}$. Then H is a subgroup of G.

Definition. Let G be a group, and H a subgroup of G . A left coset of H, generated by $g \in G$, is

$$gH = \{ gh \mid h \in H \} .$$

Proposition 3.1 Let H be a subgroup of G, and let gH be a left coset. Then there is a 1 - 1 correspondence between the elements of H and the elements of gH. (In particular, if H is finite, they have the same number of elements.)

Proof : Map $H \longrightarrow gH$ by $h \longmapsto gh$. By definition of gH, this map is onto. So suppose $h_1, h_2 \in H$ have the same image. Then

$$gh_1 = gh_2.$$

Multiplying on the left by g^{-1}, we deduce $h_1 = h_2$.

Corollary 3.2 Let G be a finite group, and let H be a subgroup. Then

$$\#(G) = \#(H) \cdot (\text{number of left cosets of } H).$$

Proof : Indeed, all the left cosets of H have the same number of elements as H, by the proposition. If $g \in G$, then $g \in gH$, since $g = g \cdot 1$, and $1 \in H$. Thus G is the union of the left cosets of H. Finally, note that two cosets gH, and $g'H$ are either equal, or disjoint. Indeed, suppose gH and $g'H$ have an element in common, namely x.

$$x = gh = g'h'.$$

Multiplying on the right by h^{-1}, we have $g = g'h'h^{-1} \in g'H$. Hence for any $y \in gH$, $y = gh'' = g'h'h^{-1}h'' \in g'H$, so $gH \subseteq g'H$. By symmetry we have the opposite inclusion, so they are equal.

The result follows immediately.

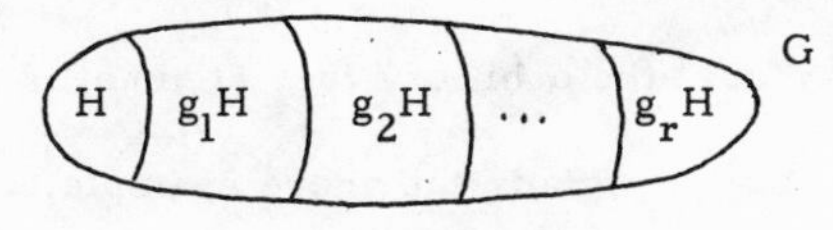

Example . Let S be a finite set, and let G be a subgroup of the group Perm S of permutations of S. Let $x \in S$, and let H be the subgroup of G leaving x fixed :

$$H = \{ g \in G \mid g(x) = x \}.$$

Let $g \in G$, and suppose $g(x) = y$. Then for any $g' \in gH$, $g'(x) = y$. Indeed, $g' = gh$ for some $h \in H$, so

$$g'(x) = gh(x) = g(x) = y.$$

Conversely, let $g'' \in G$ be some element such that $g''(x) = y$. Then

$$g^{-1}g''(x) = g^{-1}(y) = x,$$

so

$$g^{-1}g'' \in H,$$

and

$$g'' = g \cdot g^{-1}g'' \in gH.$$

Thus

$$gH = \{g' \in G \mid g'(x) = y\}.$$

It follows that the number of left cosets of H is equal to the number of point in the <u>orbit</u> of x under G. The orbit of x is the set of points $y \in S$ such that $y = g(x)$ for some $g \in G$. So we conclude

$$\#(G) = \#(H) \cdot \#(\text{orbit } x)$$

<u>Definition :</u> A group $G \subseteq \text{Perm } S$ of permutations of a set S is <u>transitive</u> if the orbit of some element is the whole of S. It follows that the orbit of every element is all of S.

So in the above example, if G is transitive,

$$\#(G) = \#(H) \cdot \#(S) .$$

<u>Corollary 3.3</u> Let S be a set with n elements, and let $G = \text{Perm } S$. Then $\#(G) = n!$

Proof: By induction on n. If $n = 1$, there is only the identity permutation, so $\#(G) = 1$. So let S have $n + 1$ elements, and let $x \in S$. Let H be the subgroup of permutations leaving x fixed. G is transitive, since one can permute x with any other element of S. Hence

$$\#(G) = \#(H) \cdot \#(S) = (n + 1) \cdot \#(H).$$

But H is just the group of permutations of the remaining n elements of S, so $\#(H) = n!$ by the induction hypothesis. Hence

$$\#(G) = (n + 1)!$$

q. e. d.

Later in the course, we will have much to do with the group of automorphisms of a projective plane, and certain of its subgroups. In particular, we will show that the axiom P 5 ("Desargues theorem") is equivalent to the statement that the group of automorphisms is "large enough", in a sense which will be made precise later. For the moment, we will content ourselves with calculating the automorphisms of a few simple configurations.

Automorphisms of the Projective Plane of Seven Points.

Call the plane π. Name its seven points A, B, C, D, P, Q, R (this suggests how it could be obtained by completing the affine plane of four points.) Then its lines are as shown.

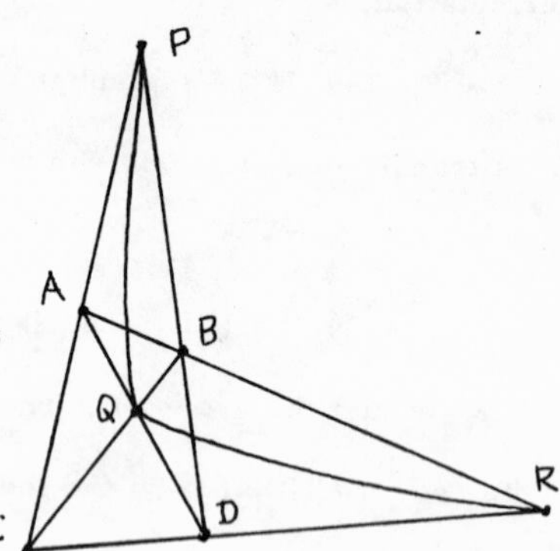

Proposition 3.4

$G = \text{Aut}\ \pi$ is transitive.

Proof: We will write down some elements of G explicitly.

$$a = (AC)(BD)$$

for example. This notation means " interchange A and C, and interchange B and D ". More generally a symbol

$$(A_1A_2, \ldots, A_r)$$

means "send A_1 to A_2, A_2 to $A_3, \ldots, A_{r-1}$ to A_r, and A_r to A_1". Multiplication of two such symbols is defined by performing the one on the right first, then the next on the right, and so on.

$$b = (AB)(CD)$$

Thus we see already that A can be sent to B or to C. We calculate

$$ab = (AC)(BD)(AB)(CD) = (AD)(BC)$$

$$ba = (AB)(CD)(AC)(BD) = (AD)(BC) = ab.$$

Thus we can also send A to D.

Another automorphism is

$$c = (BQ)(DR)$$

Since the orbit of A already contains B, C, D, we see that it also contains Q and R. Finally

$$d = (PA)(BQ)$$

shows that the orbit of A is all of π, so G is transitive.

Proposition 3.5 Let $H \subseteq G$ be the subgroup of automorphisms of π leaving P fixed. Then H is transitive on the set $\pi - \{P\}$.

Proof : Note that a, b, c above are all in H, so that the orbit of A under H is $\{A, B, C, D, Q, R\} = \pi - \{P\}$.

Theorem 3.6 Given two sets A_1, A_2, A_3 and A_1', A_2', A_3' of three non-collinear points of π, there is one and only one automorphism of π which sends A_1 to A_1' , A_2 to A_2', and A_3 to A_3' . The number of elements in G is $7 \cdot 6 \cdot 4 = 168$.

Proof : We carry the above analysis one step farther as follows. Let $K \subseteq H$ be the subgroup leaving Q fixed. Therefore since elements of K leave P and Q fixed, they also leave R fixed. K is transitive on the set $\{A, B, C, D\}$, since $a, b \in K$. On the other hand, an element of K is uniquely determined by where it sends the point A, as one sees easily. Hence K is just the group consisting of the four elements. 1, a, b, ab. We conclude from the previous discussion that

$$\#(G) = \#(H) \cdot \#(\pi)$$

$$\#(H) = \#(K) \cdot \#(\pi - \{P\}),$$

whence

$$\#(G) = 7 \cdot 6 \cdot 4 = 168.$$

The first statement of the theorem follows from the previous statements, but it is a little tricky. We do it in three steps.

1) Since G is transitive, we can find $g \in G$ such that

$$g(A_1) = A_1'$$

2) Again since G is transitive, we can find $g_1 \in G$ such that

$$g_1(P) = A_1.$$

Then

$$gg_1(P) = A_1'.$$

We have supposed that $A_1 \neq A_2$, and $A_1' \neq A_2'$. Thus

$$g_1^{-1}(A_2) \text{ and } (gg_1)^{-1}(A_2')$$

are distinct from P. But H is transitive on $\pi - \{P\}$, so there is an element $h \in H$ such that

$$h(g_1^{-1}(A_2)) = (gg_1)^{-1}(A_2').$$

One checks then that

$$g' = gg_1hg_1^{-1}$$

has the property

$$g'(A_1) = A_1'$$

$$g'(A_2) = A_2'$$

3) Thus part 2) shows that any two distinct points can be sent into any two distinct points. Changing notation, we write g instead of g', so

we may assume

$$g(A_1) = A_1'$$

$$g(A_2) = A_2'.$$

Choose $g_1 \in G$ such that

$$g_1(P) = A_1$$

$$g_1(Q) = A_2,$$

by part 2) . Then since A_1, A_2, A_3 are non-collinear, and A_1', A_2', A_3' are non-collinear, we deduce that P, Q , and each of the points

$$g_1^{-1}(A_3) \text{ , } (gg_1)^{-1}(A_3')$$

are non-collinear. In other words, these last two points are in the set $\{A, B, C, D\}$. Thus there is an element $k \in K$ such that

$$k(g_1^{-1}(A_3)) = (gg_1)^{-1}(A_3').$$

One check easily that

$$g' = gg_1kg_1^{-1}$$

is the required element of G :

$$g'(A_1) = A_1'$$

$$g'(A_2) = A_2'$$

$$g'(A_3) = A_3'.$$

For the uniqueness of this element, let us count the number of triples of non-collinear points in π . The first can be chosen in 7 ways, the second in 6 ways, and the last in 4 ways. Thus there are 168 such triples. Since the order of G is 168 , there must be exactly one transformations of G sending a given triple into another such triple.

q. e. d.

Automorphisms of the Affine Plane of 9 points.

A Similar analysis of the affine plane of 9 points shows that the group of automorphisms has order $9 \cdot 8 \cdot 6 = 432$, and any three non-collinear points can be taken into any three non-collinear points by a unique element of the group.

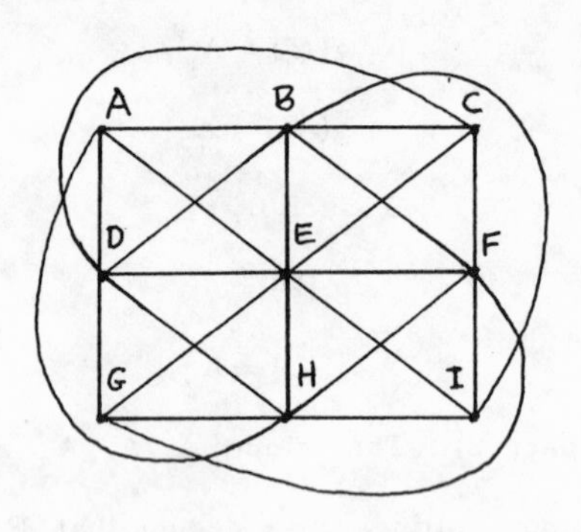

Note : In proof of theorem 3.6 , it would be sufficient to show that there is a unique automorphism sending P, Q, A into a given triple A_1, A_2, A_3 of non-collinear points. For then one can do this for each of the triples A_1, A_2, A_3, and A_1', A_2', A_3' , and compose the inverse of the first automorphism with the second. The proof thus becomes much simpler.

Automorphisms of the Real Projective Plane

Here we study another important example of the automorphisms of a projective plane. Recall that the real projective plane is defined as follows : A point is given by homogeneous coordinates (x_1, x_2, x_3) . That is, a triple of real numbers, not all zero, and with the convention that (x_1, x_2, x_3) and $(\lambda x_1, \lambda x_2, \lambda x_3)$ re present the same point, for any $\lambda \neq 0$, $\lambda \in \mathbb{R}$. A line is the set of points which satisfy an equation of the form

$$a_1 x_1 + a_2 x_2 + a_3 x_3 = 0$$

$a_i \in \mathbb{R}$, not all zero.

Brief review of matrices. An n × n matrix of real numbers is a collection of n^2 real numbers, indexed by two indices, say i, j, each of which may take values from 1 to n. Hence $A = \{a_{11}, a_{12}, \ldots, a_{21}, a_{22}, \ldots, a_{n1}, a_{n2}, \ldots, a_{nn}\}$. The matrix is usually written in a square:

$$A = (a_{ij}) = \begin{pmatrix} a_{11} & a_{12} & \cdots & a_{1n} \\ a_{21} & a_{22} & \cdots & a_{2n} \\ \vdots & & & \\ a_{n1} & a_{n2} & \cdots & a_{nn} \end{pmatrix}$$

Here the first subscript determines the row, and the second subscript determines the column.

The product of two matrices $A = (a_{ij})$ and $B = (b_{ij})$ (both of order n) is defined to be

$$A \cdot B = C$$

where $C = (c_{ij})$ and

$$c_{ij} = \sum_{k=1}^{n} a_{ik} b_{kj} .$$

$$\begin{pmatrix} a_{i1}, \ldots, a_{in} \end{pmatrix} \cdot \begin{pmatrix} b_{1j} \\ \vdots \\ b_{nj} \end{pmatrix} = \begin{pmatrix} & j \\ i & c_{ij} \end{pmatrix}$$

$$c_{ij} = a_{i1} b_{1j} + a_{i2} b_{2j} + \ldots + a_{in} b_{nj} .$$

There is also a function <u>determinant,</u> from the set of $n \times n$ matrices to $\mathbb{R}$, which is characterized by the following two properties

<u>D 1.</u> If A, B are two matrices,

$$\det(A \cdot B) = \det A \cdot \det B$$

<u>D 2.</u> For each $a \in \mathbb{R}$, let $C(a) = \begin{pmatrix} a & & & 0 \\ & 1 & & \\ & & 1 & \\ & & & \ddots \\ 0 & & & & 1 \end{pmatrix}$.

Then $\det(C(a)) = a$.

Note incidentally that the identity matrix $I = C(1)$ be haves as a multiplicative identity. One can prove the following facts :

1. $(A \cdot B) \cdot C = A \cdot (B \cdot C)$, i.e. multiplication of matrices is associative. (In general it is not commutative.)

2. A matrix A has a multiplicative inverse A^{-1} if and only if $\det A \neq 0$.

Hence the set of $n \times n$ matrices A with $\det A \neq 0$ forms a group under multiplication, denoted by $GL(n, \mathbb{R})$.

3. Let $A = (a_{ij})$ be a matrix, and consider the set of simultaneous linear equations

$$a_{11}x_1 + a_{12}x_2 + \dots + a_{1n}x_n = b_1$$

$$a_{n1}x_1 + a_{n2}x_2 + \dots + a_{nn}x_n = b_n \ .$$

If $\det A \neq 0$, then this set of equations has a solution. Conversely, if this set of equations has a solution for all possible choices of $b_1, \dots, b_n$, then $\det A \neq 0$.

For proofs on these statements, refer to any book on algebra. We will take them for granted, and use them without comment in the rest of the course. (One can prove easily that 3 follows from 1 and 2. Because to say $x_1, \ldots, x_n$ are a solution of that system of linear equations is the same as saying that

$$A \cdot \begin{pmatrix} x_1 & \\ x_2 & \\ \cdot & O \\ \cdot & \\ \cdot & \\ x_n & \end{pmatrix} = \begin{pmatrix} b_1 & \\ b_2 & \\ \cdot & O \\ \cdot & \\ \cdot & \\ b_n & \end{pmatrix} .)$$

Now let $A = (a_{ij})$ be a 3×3 matrix of real numbers, and let π be the real projective plane, with homogeneous coordinates x_1, x_2, x_3. We define a transformation T_A of π as follows : The point (x_1, x_2, x_3) goes into the point

$$T_A(x_1, x_2, x_3) = (x_1', x_2', x_3')$$

where

$$x_1' = a_{11}x_1 + a_{12}x_2 + a_{13}x_3$$
$$x_2' = a_{21}x_1 + a_{22}x_2 + a_{23}x_3$$
$$x_3' = a_{31}x_1 + a_{32}x_2 + a_{33}x_3 \quad .$$

<u>Proposition 3.7</u> If A is a 3×3 matrix of real numbers with $\det A \neq 0$, then T_A is an automorphism of the real projective plane π .

<u>Proof :</u> We must observe several things.

1) If we replace (x_1, x_2, x_3) by $(\lambda x_1, \lambda x_2, \lambda x_3)$, then (x_1', x_2', x_3') is replaced by $(\lambda x_1', \lambda x_2', \lambda x_3')$, so the mapping is well-defined. We must also check that x_1', x_2', x_3' are not all zero. Indeed,

in a matrix notation,

$$A \cdot \begin{pmatrix} x_1 \\ x_2 \\ x_3 \end{pmatrix} = \begin{pmatrix} x_1' \\ x_2' \\ x_3' \end{pmatrix}$$

where $\begin{pmatrix} x_1 \\ x_2 \\ x_3 \end{pmatrix}$ stands for the matrix

$$\begin{pmatrix} x_1 & o & o \\ x_2 & o & o \\ x_3 & o & o \end{pmatrix}.$$

But since $\det A \neq o$, A has an inverse A^{-1}, and so multiplying on the left by A^{-1}, we have

$$(\underline{x}) = A^{-1}(\underline{x}')$$

(where $(\underline{x})$ stands for the column vector $\begin{pmatrix} x_1 \\ x_2 \\ x_3 \end{pmatrix}$ etc.) So if the x_i' are all zero, the x_i are also all zero, which is impossible. Thus T_A is a well-defined map of π into π.

2) The expression $(\underline{x}) = A^{-1}(\underline{x}')$ shows that $T_{A^{-1}}$ is the inverse mapping to T_A, hence T_A must be one-to-one and surjective.

3) Finally, we must check that T_A takes lines into lines. Indeed, let

$$c_1x_1 + c_2x_2 + c_3x_3 = o \qquad (*)$$

be the equation of a line. We must find a new line, such that whenever (x_1, x_2, x_3) satisfy this equation (*), its image (x_1', x_2', x_3') lies on the new line. Let $A^{-1} = (b_{ij})$. Then we have

$$x_i = \Sigma\ b_{ij}x_j'$$

for each i. Thus if (x_1, x_2, x_3) satisfy (*) . Then (x_1', x_2', x_3') will satisfy the equation

$$c_1(\Sigma\, b_{1j} x_j') + c_2(\Sigma\, b_{2j} x_j') + c_3(\Sigma\, b_{3j} x_j') = o$$

which is

$$(\Sigma\, c_i b_{i1}) x_1' + (\Sigma\, c_i b_{i2}) x_2' + (\Sigma\, c_i b_{i3}) x_3' = o.$$

This is the equation of the required line. We have only to check that the three coefficients

$$c_j' = \Sigma\, c_i b_{ij} \qquad (**)$$

for $j = 1, 2, 3$, are not all zero. But this argument is analogous to the argument in 1) above : The equations (**) re present the fact that

$$(c_1, c_2, c_3) \cdot A^{-1} = (c_i', c_2', c_3')$$

where

$$(c_1, c_2, c_3) = \begin{pmatrix} c_1 & c_2 & c_3 \\ o & o & o \\ o & o & o \end{pmatrix}.$$

Multiplying by A on the right shows that the c_i can be expressed in terms of the c_i' . Hence if the c_i' were all zero, the c_i would be all zero, which is impossible since (*) is a line.

Hence T_A is an automorphism of π

<u>Proposition 3.8</u> Let A and A' be two 3×3 matrices with $\det A \neq o$ and $\det A' \neq o$. Then the automorphisms T_A and $T_{A'}$ of π are equal if and only if there is a real number $\lambda \neq o$, such that $A' = \lambda A$, i.e. $a'_{ij} = \lambda a_{ij}$ for all i, j .

<u>Proof :</u> Clearly if there is such a λ, $T_A = T_{A'}$, becuase the x_i' will just be changed by λ.

Conversely, suppose $T_A = T_{A'}$. We will then study the action of T_A and $T_{A'}$ on four specific points of π, namely $(1, o, o)$, $(o, 1, o)$, $(o, o, 1)$, and $(1, 1, 1)$. Let us call these points P_1, P_2, P_3, and Q respectively. Now

$$T_A(P_1) = A \cdot \begin{pmatrix} 1 \\ o \\ o \end{pmatrix} = \begin{pmatrix} a_{11} \\ a_{21} \\ a_{31} \end{pmatrix}$$

and

$$T_{A'}(P_1) = A' \cdot \begin{pmatrix} 1 \\ o \\ o \end{pmatrix} = \begin{pmatrix} a'_{11} \\ a'_{21} \\ a'_{31} \end{pmatrix} .$$

Now these two sets of coordinates are supposed to represent the same points of π, so there must exist a $\lambda \in \mathbb{R}$, $\lambda \neq o$, such that

$$a'_{11} = \lambda_1 a_{11}$$
$$a'_{21} = \lambda_1 a_{21}$$
$$a'_{31} = \lambda_1 a_{31} .$$

Similarly, applying T_A and $T_{A'}$ to the points P_2, and P_3, we find the numbers $\lambda_2 \in \mathbb{R}$ and $\lambda_3 \in \mathbb{R}$, both $\neq o$, such that

$$a'_{12} = \lambda_2 a_{12} \qquad a'_{13} = \lambda_3 a_{13}$$
$$a'_{22} = \lambda_2 a_{22} \qquad a'_{23} = \lambda_3 a_{23}$$
$$a'_{32} = \lambda_2 a_{32} \qquad a'_{33} = \lambda_3 a_{33}$$

Now apply T_A to the point Q. We find

$$A \cdot \begin{pmatrix} 1 \\ 1 \\ 1 \end{pmatrix} = \begin{pmatrix} a_{11} + a_{12} + a_{13} \\ a_{21} + a_{22} + a_{23} \\ a_{31} + a_{32} + a_{33} \end{pmatrix} .$$

Similarly for $T_{A'}$. Again, $T_A(Q) = T_{A'}(Q)$, so there is a real number $\mu \neq 0$ such that $T_{A'}(Q) = \mu \cdot T_A(Q)$. Now using all our equations, we find

$$a_{11}(\lambda_1 - \mu) + a_{12}(\lambda_2 - \mu) + a_{13}(\lambda_3 - \mu) = 0$$

$$a_{21}(\lambda_1 - \mu) + a_{22}(\lambda_2 - \mu) + a_{23}(\lambda_3 - \mu) = 0$$

$$a_{31}(\lambda_1 - \mu) + a_{32}(\lambda_2 - \mu) + a_{33}(\lambda_3 - \mu) = 0 \quad .$$

In other words, the point $(\lambda_1-\mu, \lambda_2-\mu, \lambda_3-\mu)$ Is sent into $(0,0,0)$. Hence $\lambda_1 = \lambda_2 = \lambda_3 = \mu$. (We saw this before: a triple of numbers, not all zero, cannot be sent into $(0,0,0)$ by A. Hence $\lambda_1-\mu = 0$, $\lambda_2-\mu = 0$, and $\lambda_3-\mu = 0$).

So $A' = \lambda A$, where $\lambda = \lambda_1 = \lambda_2 = \lambda_3 = \mu$, and we are done.

Definition : The projective general linear group, written $PGL(2, \mathbb{R})$ of order 2 over $\mathbb{R}$, is the group of all automorphisms of π of the form T_A for some 3×3 matrix A with $\det A \neq 0$.

Hence an element of $PGL(2, \mathbb{R})$ is represented by a 3×3 matrix $A = (a_{ij})$ of real numbers, with $\det A \neq 0$, and two matrices A, A' represent the same element of the group if and only if there is a real number $\lambda \neq 0$ such that $A' = \lambda A$.

Theorem 3.9 Let A, B, C, D and A', B', C', D' be two sets of four points, no three of which are collinear, in the real projective plane π. Then there is a unique automorphism $T \in PGL(2, \mathbb{R})$ such that $T(A) = A'$, $T(B) = B'$, $T(C) = C'$. amd $T(D) = D'$.

Proof : Let P_1, P_2, P_3, Q be the four points (1, o, o) (o, 1, o) (o, o, 1) and (1, 1, 1) considered above . Then it will be sufficient to prove the theorem in the case $A, B, C, D = P_1, P_2, P_3, Q$. Indeed, suppose we can send the quadruple P_1, P_2, P_3, Q into any other. Let φ sned it to A, B, C, D , and let ψ send it to A', B', C', D'. Then $\psi\varphi^{-1}$ sends A, B, C, D into A', B', C', D'.

Let A, B, C, D, have homogeneous coordinates (a_1, a_2, a_3) , (b_1, b_2, b_3) , (c_1, c_2, c_3) and (d_1, d_2, d_3) , respectively. Then we must find a matrix (t_{ij}) , with determinant $\neq$ o, and real numbers λ, μ, ν, ρ such that

$$T(P_1) = A \text{ i.e. } \quad \lambda a_i = t_{i1} \qquad i = 1, 2, 3$$

$$T(P_2) = B \text{ i.e. } \quad \mu b_i = t_{i2} \qquad i = 1, 2, 3$$

$$T(P_3) = C \text{ i.e. } \quad \nu c_i = t_{i3} \qquad i = 1, 2, 3$$

$$T(Q) = D \text{ i.e. } \quad \rho d_i = t_{i1} + t_{i2} + t_{i3} \qquad i = 1, 2, 3 .$$

Clearly it will be sufficient to take $\rho = 1$, and find $\lambda, \mu, \nu \neq$ o such that

$$\lambda a_1 + \mu b_1 + \nu c_1 = d_1$$

$$\lambda a_2 + \mu b_2 + \nu c_2 = d_2$$

$$\lambda a_3 + \mu b_3 + \nu c_3 = d_3 .$$

Lemma 3.10 Let A, B, C be three points in π , with coordinates (a_1, a_2, a_3) , (b_1, b_2, b_3) , (c_1, c_2, c_3) respectively. Then A, B, C, are collinear if and only if

$$\det\begin{pmatrix} a_1 & a_2 & a_3 \\ b_1 & b_2 & b_3 \\ c_1 & c_2 & c_3 \end{pmatrix} = o.$$

<u>Proof of lemma.</u> The points A, B, C are collinear if and only if there is a line, with equation say

$$h_1x_1 + h_2x_2 + h_3x_3 = o$$

h_i not all zero, such that this equations is satisfied by the coordinates of A, B, C. We have seen that the determinant of a matrix (a_{ij}) is $\neq o$ if and only if for each set of numbers (b_i), the corresponding set of linear equations (# 3. on p. 30) have a unique solution. It follows that $\det(a_{ij}) = o$ if and only if for $b_i = o$, the set of equations has a non-trivial solution, i. e. not all zero. Now our h_i are solutions of such a set of equations. Therefore they exist $\Leftrightarrow$ the determinant above is zero.

<u>Proof of theorem , continued .</u> In our case, A, B, C are non-collinear , hence by the lemma,

$$\det\begin{pmatrix} a_1 & b_1 & c_1 \\ a_2 & b_2 & c_2 \\ a_1 & b_2 & c_3 \end{pmatrix} \neq o . \quad \text{(see note below).}$$

Hence we can solve the equations above for λ, μ, ν . Now I claim λ, μ, ν, are all $\neq o$. Indeed , suppose, say $\lambda = o$. Then our equations say that

$$\mu b_1 + \nu c_1 - 1\, d_1 = o$$

$$\mu b_2 + \nu c_2 - 1\, d_2 = o$$

$$\mu b_3 + \nu c_3 - 1\, d_3 = o$$

and hence

$$\det \begin{pmatrix} b_1 & c_1 & d_1 \\ b_2 & c_2 & d_2 \\ b_3 & c_3 & d_3 \end{pmatrix} = 0$$

which is impossible, by the lemma, since B, C, D are not collinear.

<u>Note :</u> We must use the fact that the determinant of the transpose of a matrix is equal to the determinant of the matrix itself. We define the <u>transpose</u> of a matrix $A = (a_{ij})$ to be $A^T = (a_{ij})$. It is obtained by reflecting the entries of the matrix in the main diagonal. One can see easily that

$$(A \cdot B)^T = B^T \cdot A^T \ .$$

Now consider the function from the set of matrices to the real numbers given by

$$A \longrightarrow \det(A^T) \ .$$

Then this function satisfies the two conditions D 1, D 2 on p. 30, therefore it is the same as the determinant function. Hence

$$\det(A) = \det(A^T) \ .$$

So we have found λ, μ, ν all $\neq 0$ which satisfy the equations above. We define t_{ij} by the equations

$$\lambda a_i = t_{i1}$$

$$\mu b_i = t_{i2}$$

$$\nu c_i = t_{i3}$$

Then (t_{ij}) is a matrix, with determinant $\neq 0$ (again by the lemma, since A, B, C are non-collinear !), so T, given by the matrix (t_{ij}), is an element of $PGL(2, \mathbb{R})$ which sends P_1, P_2, P_3, Q to A, B, C, D.

For the uniqueness, suppose that T and T' are two elements of $PGL(2, \mathbb{R})$ which accomplish our task. Then by the proof of Proposition 3.8, the matrices (t_{ij}) and (t'_{ij}) defining T, T', differ by a scalar multiple, and hence give the same element of $PGL(2, \mathbb{R})$

q.e.d.

Our next main theorem will be that $PGL(2, \mathbb{R})$, which we know to be a subgroup of Aut π, the group of automorphisms of the real projective plane, is actually equal to it :

$$PGL(2, \mathbb{R}) = \text{Aut}\ \pi .$$

The statement and proof of this theorem will follow after some preliminary results.

<u>Definition</u> A <u>field</u> is a set F, together with two operations $+$, $\cdot$, which have the following properties.

F 1. $a + b = b + a \quad \forall\, a, b \in F$

F 2. $(a + b) + c = a + (b + c) \quad \forall\, a, b, c \in F$

F 3. $\exists\ o \in F$ such that $a + o = o + a = a \quad \forall\, a \in F$

F 4. $\forall\, a \in F$, $\exists\ -a \in F$ such that $a + (-a) = o$.

In other words, F is an abelian group under addition.

F 5. $ab = ba \quad \forall\, a, b \in F$.

F 6. $a(bc) = (ab)c \quad \forall\, a, b, c \in F$

F 7. $\exists\ 1 \in F$ such that $a \cdot 1 = a \quad \forall\, a \in F$

F 8. $\forall\, a \neq o,\ a \in F,\ \exists\, a^{-1}$ such that $a \cdot a^{-1} = 1$.

F 9. $a(b + c) = ab + ac \qquad \forall\, a, b, c \in F$.

So the non-zero elements form a group under multiplication (it is normal to assume also $o \neq 1$.) .

Definition : If F is a field , an automorphism of F is a 1 - 1 mapping σ of F onto F, written $a \longrightarrow a^{\sigma}$, such that

$$(a + b)^{\sigma} = a^{\sigma} + b^{\sigma}$$
$$(ab)^{\sigma} = a^{\sigma} b^{\sigma}$$

for all $a, b \in F$. (It follows that $o^{\sigma} = o$, $1^{\sigma} = 1$.)

Proposition 3.11 Let φ be any automorphism of the real projective plane which leaves fixed the points $P_1 = (1, o, o)$, $P_2 = (o, 1, o)$, $P_3 = (o, o, 1)$ and $Q = (1, 1, 1)$. (Note we do not assume that φ can be given by a matrix.)Then there is an automorphism σ of the field of real numbers, such that

$$\varphi(x_1, x_2, x_3) = (x_1^{\sigma}, x_2^{\sigma}, x_3^{\sigma})$$

for each point (x_1, x_2, x_3) of π.

Proof : We note that φ must leave the line $x_3 = o$ fixed since it contains P_2 and P_1. We will take this line as the line at infinity, and consider the affine plane $x_3 \neq o$. $A = \pi - \{x_3 = o\}$.

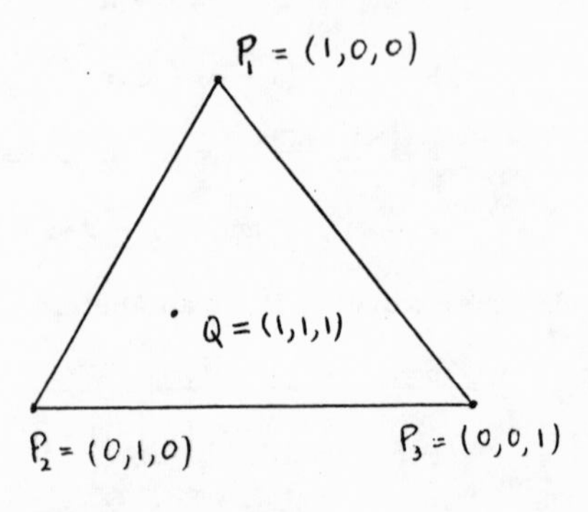

Our automorphism φ then sends A into itself, and so is an automorphism of the affine plane. We will use affine coordinates

$$x = x_1/x_3$$
$$y = x_2/x_3.$$

Since φ leaves fixed P_1 and P_2, it will send horizontal lines into horizontal lines, vertical lines into vertical lines. Besides that, it leaves fixed $P_3 = (o, o)$ and $Q = (1, 1)$, hence it leaves fixed the X-axis and the Y-axis.

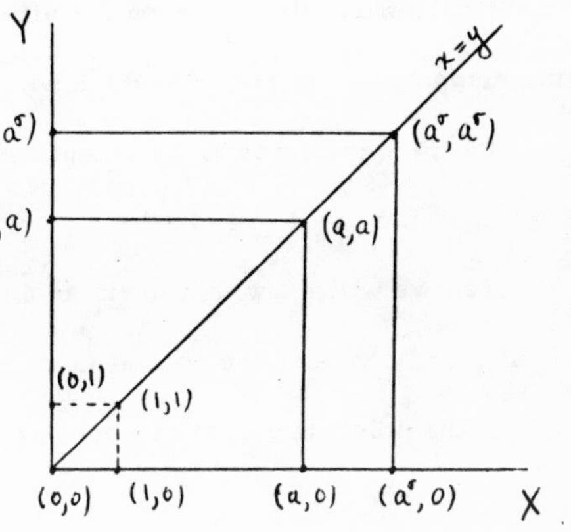

Let (a, o) be a point on the X-axis. Then $\varphi(a, o)$ is also on the X-axis, so it can be written as (a^σ, o) for a suitable element $a^\sigma \in \mathbb{R}$. Thus we define a mapping

$$\sigma : \mathbb{R} \longrightarrow \mathbb{R} ,$$

and we see immediately that $o^\sigma = o$ and $1^\sigma = 1$.

The line $x = y$ is sent into itself, because P_3 and Q are fixed. Vertical lines go into vertical lines. Hence the point

$$(a, a) = (\text{line } x = y) \cap (\text{line } x = a)$$

is sent into

$$(a^\sigma, a^\sigma) = (\text{line } x = y) \cap (\text{line } x = a)$$

Similarly, horizontal lines go into horizontal lines, and the Y-axis goes into itself, so we deduce that

$$\varphi(o, a) = (o, a^{\sigma}) .$$

Finally, if (a, b) is any point, we deduce by drawing the lines $x = a$ and $y = b$ that

$$\varphi(a, b) = (a^{\sigma}, b^{\sigma}) .$$

Hence the action of φ on the affine plane is completely expressed by the mapping $\sigma : \mathbb{R} \longrightarrow \mathbb{R}$ which we have constructed.

By the way, since φ is an automorphism of A, it must send the X-axis onto itself in a 1 - 1 manner, so σ is one - to - one and onto.

Now we will show that σ is an automorphism of $\mathbb{R}$. Let $a, b \in \mathbb{R}$, and consider the points (a, o), (b, o) on the X-axis. We can construct the point (a + b, o) geometrically as follows :

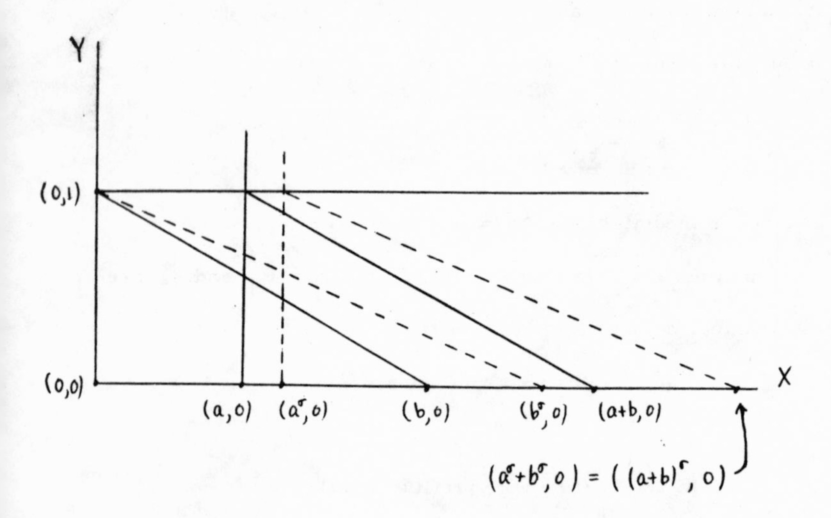

1. Draw the line $y = 1$

2. Draw $x = a$

3. Get $(a, 1)$ by intersection of 1, 2.

4. Draw the line joining $(o, 1)$ and (b, o).

5. Draw the line parallel to 4. through $(a, 1)$.

6. Intersect 5. with the X-axis.

Now φ sends the line $y = 1$ into itself; it sends $x = a$ into $x = a^\sigma$, and it sends (b, o) into (b^σ, o). It preserves joins and intersections, and parallelism. Hence φ also sends $(a + b, o)$ into $(a^\sigma + b^\sigma, o)$. Therefore

$$(a + b)^\sigma = a^\sigma + b^\sigma .$$

By another construction, we can obtain the point (ab, o) geometrically from the points (a, o) and (b, o).

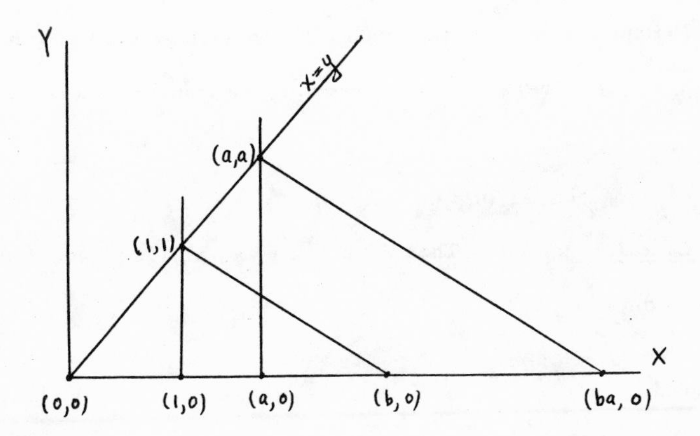

1. Draw $x = a$.

2. Intersect with $x = y$ to obtain (a, a) .

3. Join $(1, 1)$ to (b, o).

4. Draw a line parallel to 3. through (a, a) .

5. Intersect 4. with the X-axis.

Since φ leaves $(1, 1)$ fixed, we see similarly by this construction that

$$(ab)^\sigma = a^\sigma b^\sigma .$$

Hence σ is an automorphism of the field of real numbers.

Now we return to the projective plane π , and study the effect of φ on a point with homogeneous coordinates (x_1, x_2, x_3) .

Case 1.: If $x_3 = o$, we write this point as the intersection of the line $x_3 = o$ (which is left fixed by φ) and the line joining $(o, o, 1)$ with $(x_1, x_2, 1)$. Now this latter point is in A, and has affine coordinates (x_1, x_2) . Hence φ of it is (x_1^σ, x_2^σ) , whose homogeneous coordinates are $(x_1^\sigma, x_2^\sigma, 1)$. Therefore by intersecting the transformed lines, we find

$$\varphi(x_1, x_2, o) = (x_1^\sigma, x_2^\sigma, o) .$$

Case 2. $x_3 \neq o$. Then the point (x_1, x_2, x_3) is in A , and has affine coordinates

$$x = x_1/x_3$$

$$y = x_2/x_3.$$

So $\varphi(x, y) = (x^\sigma, y^\sigma) = (x_1^\sigma/x_3^\sigma, x_2^\sigma/x_3^\sigma)$. This last equations because

σ is an automorphism , so takes quotients into quotients. Therefore $\varphi(x, y)$ has homogeneous coordinates $(x_1^\sigma, x_2^\sigma, x_3^\sigma)$ and we are done.

q. e. d.

Proposition 3.12 The only automorphism of the field of real numbers is the identity automorphism.

Proof : Let σ be an automorphism of the real numbers. We proceed in several steps.

1) $1^\sigma = 1$. $(a + b)^\sigma = a^\sigma + b^\sigma$. Hence , by induction, we can prove that $n^\sigma = n$ for any positive integer n.

2) $n + (-n) = o$, so $n^\sigma + (-n)^\sigma = o$, so $(-n)^\sigma = -n$.
Hence σ leaves all the integers fixed.

3) If $b \neq o$, $(a/b)^\sigma = a^\sigma/b^\sigma$. Hence σ leaves all the rational numbers fixed.

4) If $x \in \mathbb{R}$, then $x > o$ if and only if there is an $a \neq o$ such that $x = a^2$. Then $x^\sigma = (a^\sigma)^2$, so $x > o \Rightarrow x^\sigma > o$. Conversely if $x^\sigma > o$, $x^\sigma = b^2$, so $x = (x^\sigma)^{\sigma^{-1}} = (b^{\sigma^{-1}})^2$, because the inverse of σ is also an automorphism. Hence $x > o \Leftrightarrow x^\sigma > o$. Therefore also $x < y \Leftrightarrow x^\sigma < y^\sigma$.

5) Let $\{a_n\}$ be a sequence of real numbers, and let a be a real number. Then the sequence $\{a_n\}$ converges to $a \Leftrightarrow \{a_n^\sigma\}$ converges to a^σ. Indeed, this says $\forall \in > o$, $\exists N$ such that $n > N \Rightarrow |a_n - a| < \in$. Using the previous results, this is equivalent to

$|a_n^\sigma - a^\sigma| < \epsilon^\sigma$. Furthermore, it is sufficient to consider rational $\epsilon > o$ in the definition, so $\epsilon^\sigma = \epsilon$ if ϵ is a rational number. So the two conditions are equivalent.

6). If $a \in \mathbb{R}$ is any real number, we can find a sequence of rational numbers $q_n \in \mathbb{Q}$, which converges to a. Then $q_n^\sigma = q_n$, q_n^σ converges to a^σ, and so $a = a^\sigma$, by the uniqueness of the limit.

Thus σ is the identity.

q.e.d.

Theorem 3.13 $PGL(2, \mathbb{R}) = \text{Aut}\, \pi$.

Proof : It is sufficient to show that any $\varphi \in \text{Aut}\, \pi$ is already in $PGL(2, \mathbb{R})$. Let $\varphi \in \text{Aut}\, \pi$. Let $\varphi(P_1) = A$, $\varphi(P_2) = B$, $\varphi(P_3) = C$, $\varphi(Q) = D$. Choose a $T \in PGL(2, \mathbb{R})$ such that $T(P_1) = A$, $T(P_2) = B$, $T(P_3) = C$, $T(Q) = D$. (Possible by theorem 3.9). Then $T^{-1}\varphi$ is an automorphism of π which leaves P_1, P_2, P_3, Q fixed. Hence by Proposition 3.10 it can be written

$$(x_1, x_2, x_3) \longrightarrow (x_1^\sigma, x_2^\sigma, x_3^\sigma)$$

for some automorphism σ of $\mathbb{R}$. But by the last Proposition σ is the identity, so $T^{-1}\varphi$ is the identity, so $\varphi = T \in PGL(2, \mathbb{R})$.

q.e.d.

Note that specific properties of the real numbers entered only into Proposition 3.11. The rest of the argument would have been valid over an arbitrary field. In fact, we will study this more general situation in Chapter 6.

CHAPTER 4. ELEMENTARY SYNTHETIC PROJECTIVE GEOMETRY

We will now study the properties of a projective plane which we can deduce from the axioms P 1 - P 4 (and occasionally P 5, P 6, P 7 to be defined)

Proposition 4.1 Let π be a projective plane. Let π^* be the set of lines in π, and define a line* in π^* to be a pencil of lines in π. (A pencil of lines is the set of all lines passing through some fixed point.) Then π^* is a projective plane, called the dual projective plane of π. Furthermore, if π satisfies P 5, so does π^*.

Proof : We must verify the asioms P 1 - P 4 for π^*, and we will call them P 1*, ... P 4* to distinguish them from P 1 - P 4. Also P 5 $\Rightarrow$ P 5*.

P 1*. If P*, Q* are two distinct points* of π^*, then there is a unique line* of π^* containing P* and Q*. If we translate this statement into π, it says, if l, m are two distinct lines of π, then there is a unique pencil of lines containing l, m, i.e. l, m have a unique point in common. This follows from P 1 and P 2.

P 2* If l* and m* are two lines* in π^*, they have at least one point* in common. In π, this says that two pencils of lines have at least one line in common, which follows from P 1.

P 3* There are three non-collinear points* in π^*. This says there are three non-concurrent lines in π. (We say three or more lines are <u>concurrent</u> if they all pass through some point, i.e. if they are contained in a pencil of lines.) By P 3 there are three non-collinear points A, B, C. Then one sees easily that the lines AB, AC, BC are not concurrent.

P 4* Every line* in π^* has at least three points*.

This says that every pencil in π has at least three lines. Let the pencil be centered at P, and let l be some line not passing through P. Then by P 4, l has at least three points A, B, C. Hence the pencil of lines through P has at least three lines $a = PA$, $b = PB$, $c = PC$.

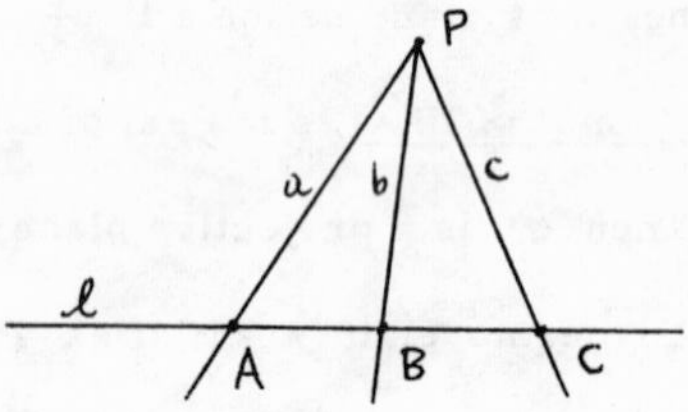

Now we will assume P 5, Desargues axiom, and we wish to prove

P 5* Let O*, A*, B*, C*, A*', B*', C*' be seven distinct points* of π^*, such that O*A*A'*, O*B*B'*, O*C*C'* are collinear, and A*, B*, C*; A'*, B'*, C'* are not collinear. Then the points*

$$P^* = A^*B^* \cdot A'^*B'^*$$

$$Q^* = A^*C^* \cdot A'^*C'^*$$

$$R^* = B^*C^* \cdot B'^*C'^*$$

are collinear.

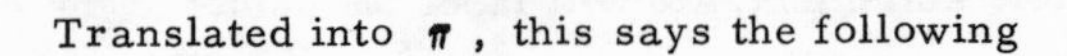

Translated into π , this says the following

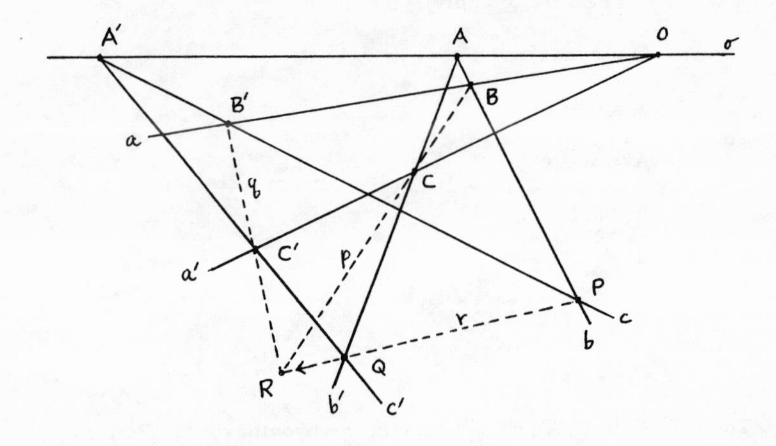

Let o, a, b, c, a', b', c' be seven lines, such that o, a, a' ; o, b, b' ; o, c, c' are concurrent, and such that a, b, c ; a', b', c' are not concurrent. Then the lines

$$p = (a.b) \cup (a'.b')$$

$$q = (a.c) \cup (a'.c')$$

$$r = (b.c) \cup (b'.c')$$

(where $\cup$ denotes the line joining two points, and . denotes the intersection of two lines) are concurrent.

To prove this statement, we will label the points of the diagram in such a way as to be able to apply P 5. So let

$$O = o.a.a'$$

$$A = o.b.b'$$

$$A' = o.c.c'$$

$$B = a.b$$

$$B' = a.c$$

$$C = a'.b'$$

$$C' = a'.c'$$

Then O, A, B, C, A', B', C' satisfy the hypothesis of P 5, so we conlude that

$$P = AB.A'B' = b.c$$

$$Q = AC.A'C' = b'.c'$$

$$R = BC.B'C' = p.q$$

are collinear. But $PQ = r$, so this says that p, q, r are concurrent.

q. e. d.

Corollary 4.2 (Principle of Duality). Let S be any statement about a projective plane π, which can be proved from the axioms P 1 - P 4 (respectively P 1 - P 5). Then the "dual" statement S*, obtained from S by interchanging the words

point <——> line

lies on <——> passes through

collinear <——> concurrent.

intersection <——> join etc.

can also be proved from the axioms P 1 - P 4 (respectively P 1 - P 5).

<u>Proof</u> : Indeed, S* is just the statement S applied to the dual projective plane π^* , hence it follows from P 1* - P 4* (resp. P 1* - P 5*) . But these in turn follow from P 1 - P 4 (respectively P 1 - P 5) , as we have just shown.

<u>Remarks</u> : 1. There is a natural map $\pi \longrightarrow \pi^{**}$, obtained by sending a point P of π into the pencil of lines through P , which is a point of π^{**} . One can see easily that this is an <u>isomorphism</u> of the projective plane π with the projective plane π^{**} .

2. However, the plane π^* need not be isomorphic to the plane π . I believe one of the non-Desarguesian finite projective planes of order 9 (10 points on a line) will give an example of this.

<u>Definition</u> : A <u>complete quadrangle</u> is the configuration of seven points and six lines obtained by taking four points A, B, C, D, no three of which are collinear , drawing all six lines connecting them , and then taking the intersection of opposite sides :

$$P = AB \cdot CD$$

$$Q = AC \cdot BD$$

$$R = AD \cdot BC$$

The points P, Q, R are called <u>diagonal points</u> of the complete quadrangle.

It may happen that the diagonal points P, Q, R of a complete quadrangle are collinear (as for example in the projective plane of seven points) . However, this never happens in the real projective plane

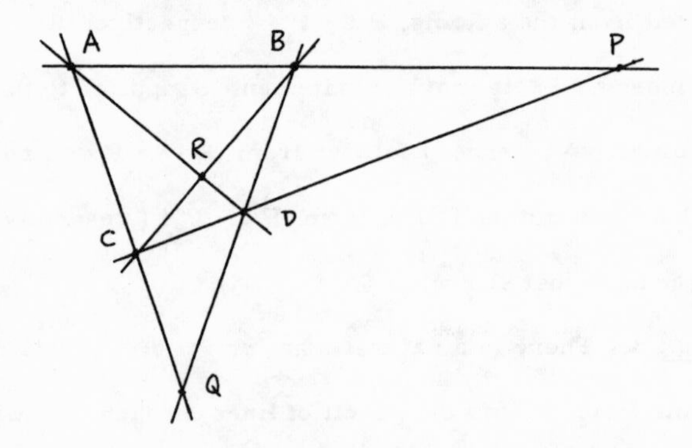

(as we will see below), and in general, it is to be regarded as a pathological phenomenon, hence we will make an axiom saying this should not happen.

P 7. (Fano's Axiom). The diagonal points of a complete quadrangle are never collinear.

Proposition 4.3 The real projective plane satisfies P 7.

Proof: Let A, B, C, D be the vertices of a complete quadrangle. Then no three of them are collinear, so we can find an automorphism T of the real projective plane π which carries A, B, C, D into the points (o, o, 1) (1, o, o) (o, 1, o), (1, 1, 1) respectively (by theorem 3.9).

Hence it will be sufficient to show that the diagonal points of this complete quadrangle are not collinear. They are (1, o, 1), (1, 1, o), and (o, 1, 1). To see if they are collinear,

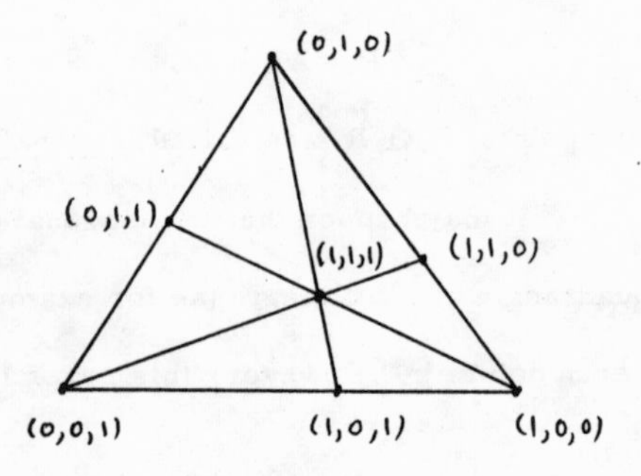

we apply lemma 3.10, and calculate determinant

$$\det\begin{pmatrix} 1 & o & 1 \\ 1 & 1 & o \\ o & 1 & 1 \end{pmatrix} = 2$$

Since $2 \neq o$, we conclude that the points are not collinear.

<u>Proposition 4.4</u> P 7 in a projective plane π implies P 7* in π^*, hence the principle of duality also applies in regard to consequences of P 7.

<u>Proof :</u> P 7*, translated into the language of π, says the following : The diagonal lines of a complete quadrilateral are never concurrent. This statement requires some explanation :

<u>Definition :</u> A <u>complete quadrilateral</u> is the configuration of seven lines and six points, obtained by taking four lines a, b, c, d, no three of which are concurrent, their six points of intersection, and the three lines

$$p = (a.b) \cup (c.d)$$

$$q = (a.c) \cup (b.d)$$

$$r = (a,d) \cup (b.c)$$

joining opposite pairs of points. These lines p, q, r are called the <u>diagonal lines</u> of the complete quadrilateral.

To prove P 7*, let a, b, c, d be a complete quadrilateral, and suppose that the three diagonal lines p, q, r were concurrent. Then this would show that the diagonal points of the complete quadrangle

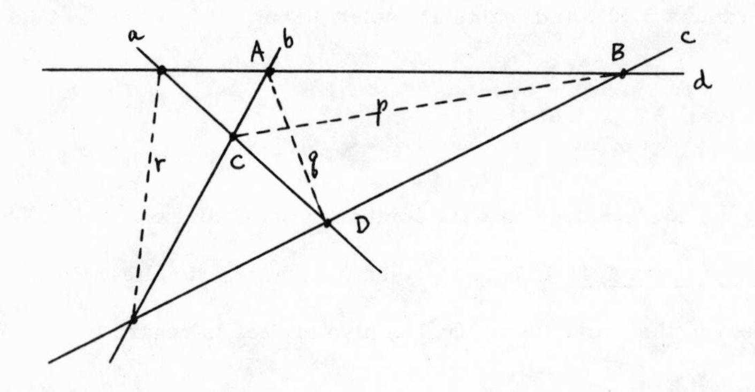

ABCD , where

$$A = b.d$$

$$B = c.d$$

$$C = a.b$$

$$D = a.c ,$$

were collinear, which contradicts P 7 . ✱. Hence P 7* is true .

<u>Remark :</u> The astute reader will have noticed that the definition of a complete quadrilateral is the "dual" of the definition of a complete quadrangle. In general, I expect from now on that the reader will construct for himself the duals of all definitions, theorems, and proofs.

Harmonic Points.

Definition . An ordered quadruple of distinct points A, B, C, D on a line is called a harmonic quadruple if there is a complete quadrangle X, Y, Z, W such that A and B are diagonal points of the complete quadrangle (say

$$A = XY \; . \; ZW$$

$$B = XZ \; . \; YW \;)$$

and C, D lie on the remaining two sides of the quadrangle (say $C \in XW$ and $D \in YZ$).

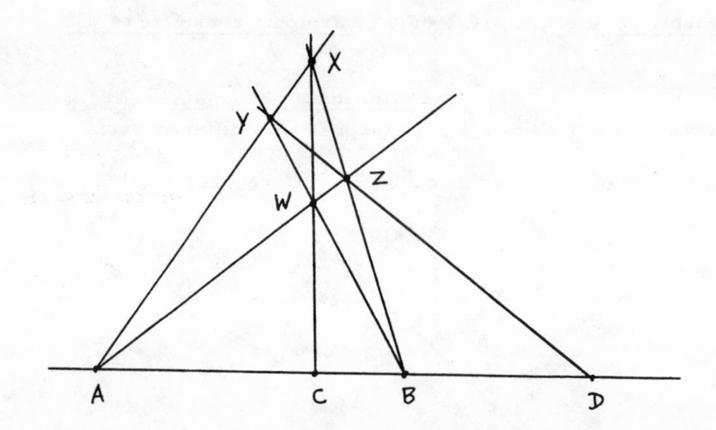

In symbols , we write H(AB, CD) if A, B, C, D form a harmonic quadruple.

Note that if ABCD is a harmonic quadruple, then the fact that A, B, C, D are distinct implies that the diagonal points of a defining quadrangle XYZW are not collinear. In fact , the notion of 4 harmonic points does not make much sense unless Fano's Axiom P 7 is satisfied, hence we will always assume this when we speak of harmonic points.

Proposition 4.5 : $H(AB, CD) \Leftrightarrow H(BA, CD) \Leftrightarrow H(AB, DC) \Leftrightarrow H(BA, DC)$.

Proof : This follows immediately from the definition, since A and B play symmetrical roles, and C and D play symmetrical roles. In fact, once could permute X, Y, Z, W to make the notation coincide with the definition of H(BA, CD) etc.

Proposition 4.6 Let A, B, C be three distinct points on a line. Then (assuming P 7) , there is a point D such that H(AB, CD) . Furthermore (assuming P 5) , this point D is unique . D is called the fourth harmonic point of A, B, C , or the harmonic conjugate of C with respect to A and B .

Proof : Draw two lines l , m through A, different from the line ABC . Draw a line n through C , different from ABC .

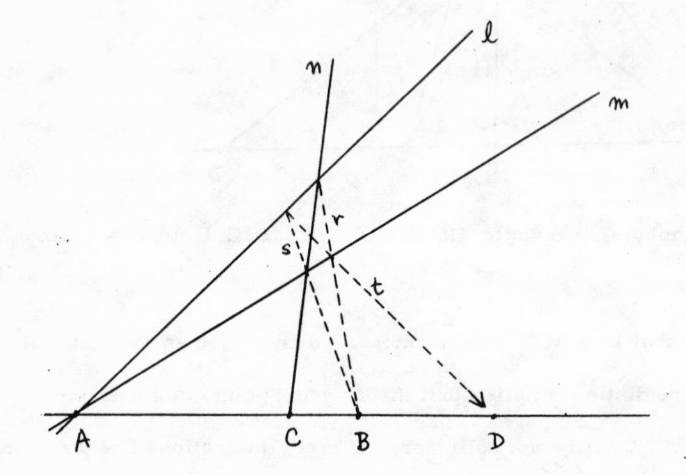

Then join B to l.n , and join B to m.n . Call these lines r , s respectively. Then join r.m and s.l to form a line t . Let t intersect ABC at D . Then by P 7 we see that D is distinct from A, B, C . Hence by construction we have H(AB, CD).

Now we assume P 5 , and will prove the uniqueness of the fourth harmonic point. Given A, B, C construct D as above. Suppose D' is another point such that H(AB, CD') . Then by definition, there is a complete quadrangle XYZW such that

$$A = XY.ZW$$

$$B = XZ.YW$$

$$C \in XW$$

$$D' \in YZ .$$

Call l' = AX , m' = AZ , and n' = CX . Then we see that the above construction, applied to l', m', n', will give D' .

Thus it is sufficient to show that our construction of D is independent of the choice of l, m, n . We do this in three steps, by showing that if we vary one of l, m, n, the point D remains the same.

Step 1. If we replace l by a line l', we get the same D.

Let D be defined by l, m, n as above, and label the resulting complete quadrangle XYZW. Let l' be another line through A, distinct from m, and label the quadrangle abtained from l', m, n , X'Y'Z'W' . (Note the point W = m.n belongs to both quadrangles.)

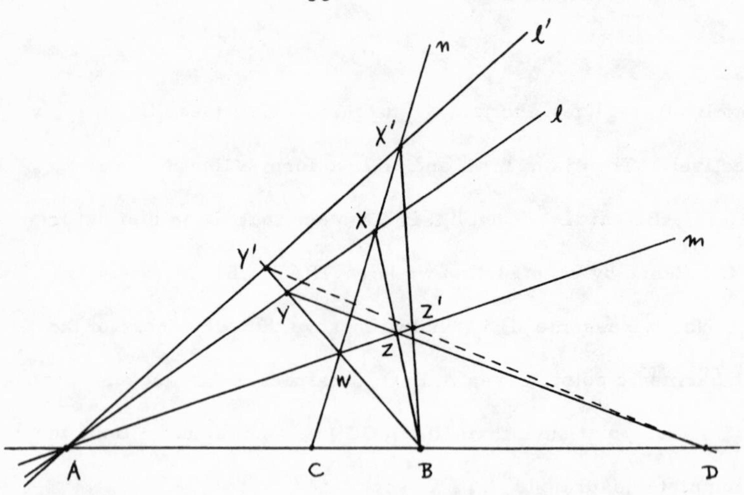

We must show that the line Y'Z' passes through D , i.e. that (Y'Z') . (ABC) = D. Indeed, observe that the two triangles XYZ and X'Y'Z' are perspective from W. Two pairs of corresponding sides meet in A and B respectively :

$$A = XY.X'Y'$$

$$B = XZ.X'Z' .$$

Hence by P 5, the third pair of corresponding sides, namely YZ and Y'Z' , must meet on AB, which is what we wanted to prove.

<u>Step 2.</u> If we replace m by m' , we get the same D. The proof in this case is identical with that of Step 1. , interchanging the roles of l and m.

<u>Step 3.</u> If we replace n by n' we get the same D.

The proof in this case is more difficult , since all four points of the corresponding complete quadrangle change. So let XYZW be the quadrangle formed by l, m, n, which defines D. Let X'Y'Z'W'

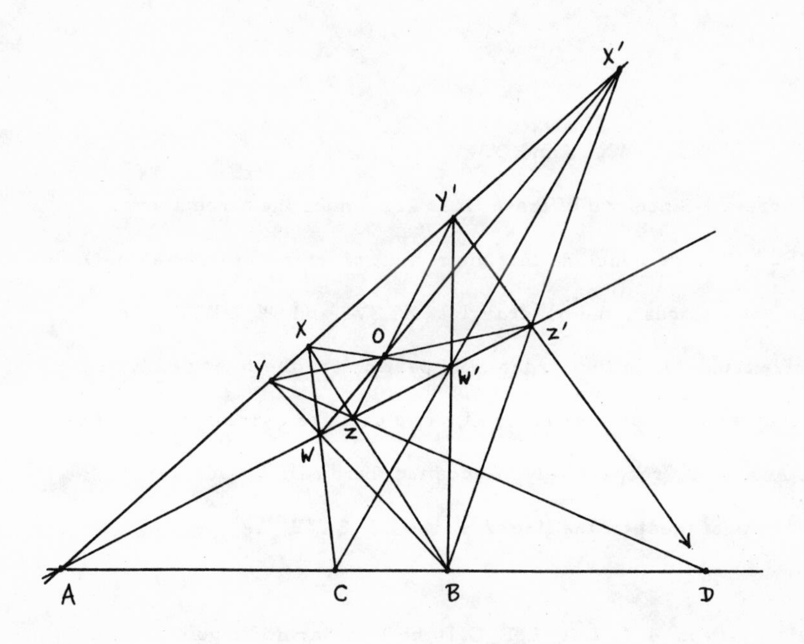

be the quadrangle formed by l, m, n' . We must show that Y'Z' also meets ABC at D.

Consider the triangles XYW and W'Z'X' (in that order). Corresponding sides meet in A, B, C, respectively, which are collinear, hence by P 5* , the two triangles must be perspective from some point O . In other words, the lines

XW', YZ', and WX'

all meet in a point O.

Similarly, by considering the triangles ZWX and Y'X'W' (in that order), and applying P 5* once more, we deduce that the lines

$$ZY', \; WX', \text{ and } XW'$$

are concurrent. Since two of these lines are among the three above, and $XW' \neq X'W$, we conclude that their point of intersection is also O.

In other words, the quadrangles XYZW and W'Z'Y'X' are perspective from O, in that order. In particular, the triangles XYZ and W'Z'Y' are perspective from O. Two pairs of corresponding sides meet in A and B, respectively. Hence the third pair of sides, YZ and Z'Y', must meet on the line AB, i.e. $D \in Z'Y'$.

q.e.d.

Proposition 4.7 Let A, B, C, D be four harmonic points. Then (assuming P 5) also CD, AB are four harmonic points.

Combining with proposition 4.5, we find therefore

$$\begin{array}{c} H(AB, CD) \Leftrightarrow H(BA, CD) \Leftrightarrow H(AB, DC) \Leftrightarrow H(BA, DC) \\ \Updownarrow \\ H(CD, AB) \Leftrightarrow H(DC, AB) \Leftrightarrow H(CD, BA) \Leftrightarrow H(DC, BA) . \end{array}$$

Proof: (see diagram on next page). We assume H(AB, CD), and let XYZW be a complete quadrangle as in the definition of harmonic quadruple.

Draw DX and CZ, and let them meet in U. Let $XW \cdot YZ = T$. Then XTUZ is a complete quadrangle with C, D as two of its diagonal points; B lies on XZ, so it will be sufficient to prove that TU passes through A. For then we will have H(CD, AB).

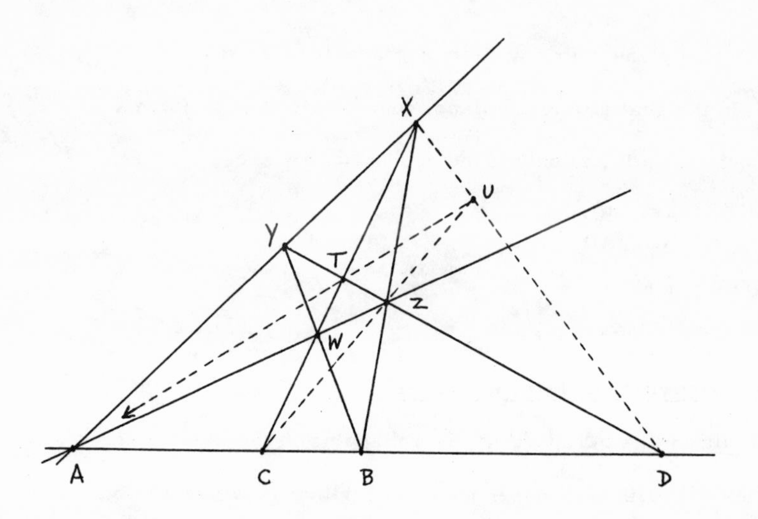

Consider the two triangles XUZ and YTW . Their corresponding sides meet in D, B, C respectively, which are collinear. Hence by P 5* , the lines joining corresponding vertices, namely

XY , TU , WZ

are concurrent, which is what we wanted to prove.

<u>Examples :</u> 1. In the projective plane of thirteen points, there are four points of any line. These four points always form a harmonic quadruple, in any order.

To prove this, it will be sufficient to show that P 7 holds in this plane. For then there will always be a fourth harmonic point to any three points, and it must be <u>the</u> fourth point on the line. We will prove this later : The plane of 13 points is the projective plane over the field of three elements, which is of characteristic 3. But P 7 holds in the projective plane over any field of characteristic $\neq 2$.

2. In the real Euclidean plane, four points ABCD form a harmonic quadruple if and only if the product of distances

$$\frac{AC}{BC} \cdot \frac{BD}{AD} = -1.$$

(see problem # 20).

Perspectivities and Projectivities

Definition : A perspectivity is a mapping of one line 1 into antother line 1' (both considered as sets of points) , which can be obtained in the following way : Let O be a point not on either 1 or 1' . For each point $A \in 1$, draw OA , and let OA meet 1' in A'. Then map $A \longrightarrow A'$. This is a perspectivity. In symbols we write

$$1 \overset{O}{\underset{\wedge}{=}} 1' ,$$

which says " 1 is mapped into 1' by a perspectivity with center at O",

or $$ABC \ldots \overset{O}{\underset{\wedge}{=}} A'B'C' \ldots ,$$

which says "the points A, B, C (of the line 1) are mapped via a perspectivity with center O into the points A', B', C' , respectively (of the line 1') ".

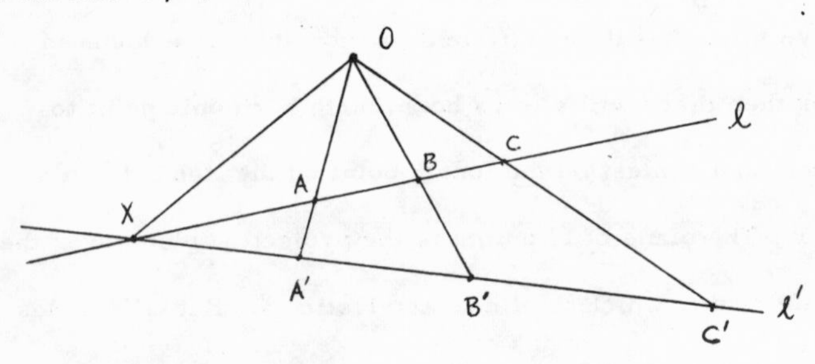

Note that a perspectivity is always one-to-one and onto, and that its inverse is also a perspectivity. Note also that if $X = l.l'$, then X (as a point of l) is sent into itself, X (as a point of l').

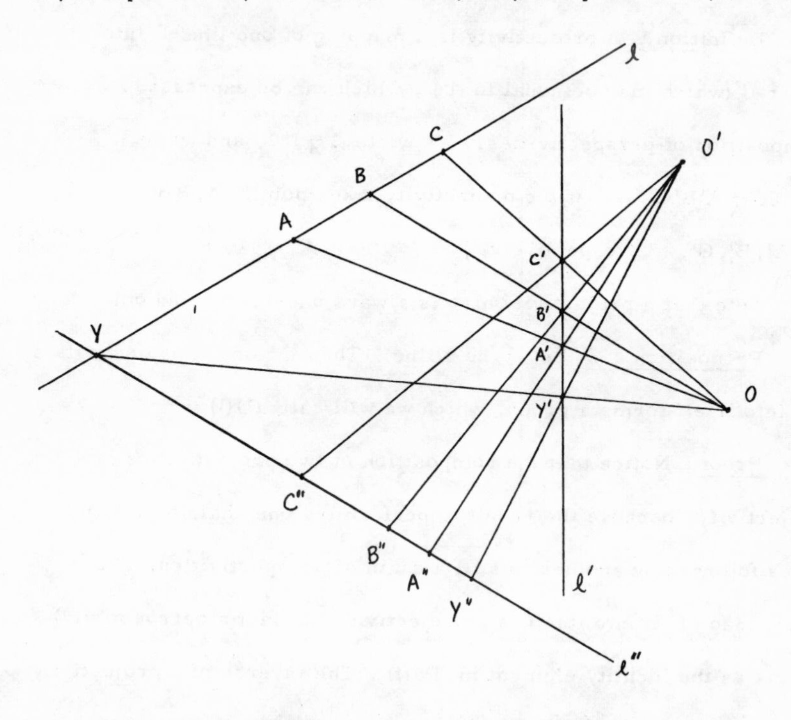

One can see easily that a composition of two or more perspectivities need not be a perspectivity. For example, in the diagram above, we have

$$l \overset{O}{\underset{\wedge}{=}} l' \overset{O'}{\underset{\wedge}{=}} l''$$

and

$$ABCY \overset{O}{\underset{\wedge}{=}} A'B'C'Y' \overset{O}{\underset{\wedge}{=}} A''B''C''Y''\ .$$

Now if the composed map, from l to l'' were a perspectivity, it would have to send $l.l'' = Y$ into itself. However, Y goes into Y'' which is different from Y. Therefore we make the following

Definition : A projectivity is a mapping of one line l into another l' (which may be equal to l), which can be expressed as a composition of perspectivities. We write $l \;\overline{\wedge}\; l'$, and write $ABC \ldots \;\overline{\wedge}\; A'B'C' \ldots$ if the projectivity takes points $A, B, C, \ldots$ into $A', B', C', \ldots$ respectively.

Note that a projectivity also is always one-to-one and onto.

Proposition 4.8 Let l be a line . Then the set of projectivities of l into itself forms a group, which we will call $PJ(l)$.

Proof : Notice that the composition of two projectivities is a projectivity, because the result of performing one chain of perspectivities followed by another is still a chain of perspectivities. The identity map of l into itself is a projectivity (in fact a perspectivity) , and acts as the identity element in $PJ(l)$. The inverse of a projectivity is projectivity, since we need only reverse the chain of perspectivities.

Naturally, we would like to study this group, and in particular, we would like to know how many times transitive it is. We will see in the following two propositions that is is three times transitive, but cannot be four times transitive.

Proposition 4.9 Let l be a line, and let $A, B, C,$ and A', B', C' be two triples of three distinct points each. Then there is a projectivity

l into itself which sends A, B, C into A', B', C'.

Proof : Let l' be a line different from l, and which does not pass through A or A'. Let O be any point not on l, l', and project A', B', C' from l to l', giving A'', B'', C'', so we have

$$A'B'C' \overset{=}{\wedge} A''B''C'' ,$$

and $A \notin l'$, $A'' \notin l$. Now it is sufficient to construct a projectivity from l to l', taking ABC into $A''B''C''$. Drop double primes, and forget the original points $A', B', C' \in l$. Thus we have the following problem :

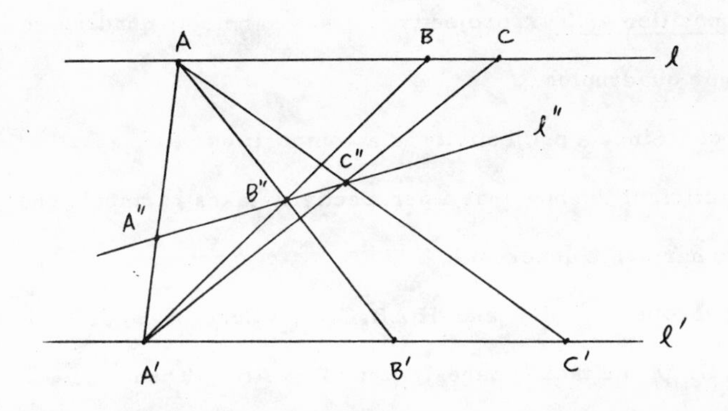

Let l, l' be two distinct lines ; let A, B, C be three distinct points on l, and let A', B', C' be three distinct points on l'; assume furthermore that $A \notin l'$, and $A' \notin l$. To construct projectivity from l to l' which carries A, B, C into A', B', C', respectively.

Draw $AA', AB', AC', A'B, A'C$, and let

$$AB' \cdot A'B = B''$$

$$AC' \cdot A'C = C'' .$$

Draw l'' joining B'' and C'', and let it meet AA' at A''. Then

$$l \overset{A'}{\underset{\wedge}{=}} l'' \overset{A}{\underset{\wedge}{=}} l'$$

sends

$$ABC \overset{A'}{\underset{\wedge}{=}} A''B''C'' \overset{A}{\underset{\wedge}{=}} A'B'C' .$$

Thus we have found the required projectivity as a composition of two perspectivities.

Proposition 4.10 A projectivity takes harmonic quadruples into harmonic quadruples.

Proof : Since a projectivity is a composition of perspectivities, it will be sufficient to show that a perspectivity takes harmonic quadruples into harmonic quadruple.

So suppose $l \overset{O}{\underset{\wedge}{=}} l'$, and $H(AB, CD)$, where $A, B, C, D \in l$. Let A', B', C', D' be their images. Let $l'' = AB'$. Then

$$l \overset{O}{\underset{\wedge}{=}} l'' \overset{O}{\underset{\wedge}{=}} l'$$

is the same mapping, so it is sufficient to consider $l \overset{O}{\underset{\wedge}{=}} l''$ and $l'' \overset{O}{\underset{\wedge}{=}} l'$ separately. Here one has the advantage that the intersection of the two lines is one of the four points considered. By relabeling, we may assume it is A in each case. So we have the following problem :

Let $l \overset{O}{\underset{\wedge}{=}} l'$, and let $A = l.l', B, C, D$ be four points on l such that $H(AB, CD)$. Prove that $H(AB', C'D')$, where B', C', D' are the images of B, C, D.

Draw BC' , and let it meet OA at X. Consider the complete quadrangle OXB'C' . Two of its diagonal points are A, B, ; C lies on the side OC' . Hence the intersection of XB' with l must be the fourth harmonic point of ABC, i. e. XB'.l = D (Here we use the unicity of the fourth harmonic point.)

Now consider the complete quadrangle OXBD. Two of its diagonal points are A and B' ; The other two sides meet l' in C' and D' . Hence H(AB', C'D').

q. e. d.

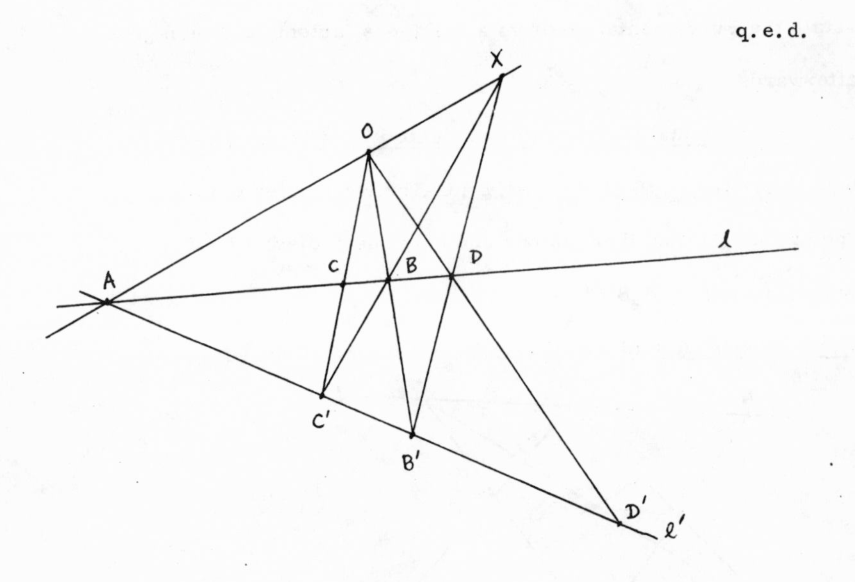

So we see that the group PJ(l) is three times transitive, but it cannot be four times transitive, because it must take quadruples of harmonic points into quadruples of harmonic points.

CHAPTER 5 . PAPPUS' AXIOM , AND THE FUNDAMENTAL THEOREM FOR PROJECTIVITIES ON A LINE.

In this chapter, we come to the "fundamental theorem", which states that there is a unique projectivity sending three points into any other three points, i.e. PJ(l) is exactly three times transitive. It turns out this theorem does not follow from the axioms P 1 - P 5 and P 7 , so we introduce P 6 , Pappus' axiom. Then we can prove the fundamental theorem, and conversely, the fundamental theorem implies P 6.
We will state the Fundamental theorem and Pappus' axiom, and then give proofs afterwards.

FT . Fundamental theorem (for projectivities on a line) .
Let l be a line. Let A, B, C and A', B', C', be two triples of three distinct points on l. Then there is one and only one projectivity of l into l such that $ABC \barwedge A'B'C'$.

P 6. Pappus' Axiom .

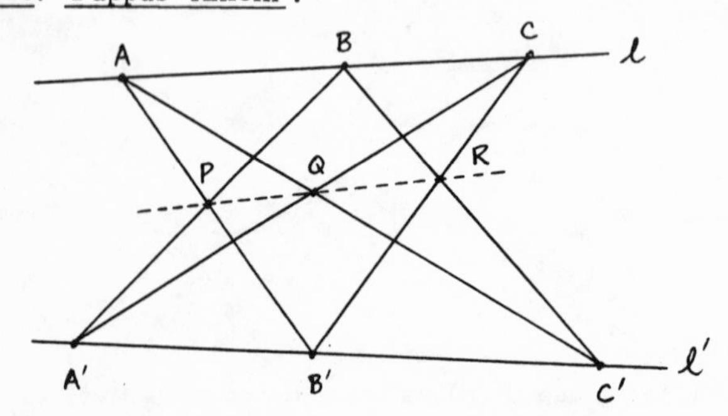

Let l and l' be two distinct lines. Let A, B, C be three distinct points on l, different from $X = l.l'$. Let A', B', C' be three distinct point on l', different from X. Define

$$P = AB'.A'B$$

$$Q = AC'.A'C$$

$$R = BC'.B'C.$$

Then P, Q, and R are collinear.

Proposition 5.1 P 6 implies the dual of Pappus' Axiom, P 6*, and so the principle of duality extends. (problem 21)

Proposition 5.2 P 6 is true in the real projective plane.

Proof : Let $l, l', A, B, C, A', B', C'$ be as in the statement, and construct P, Q, R. We take l to be the line at infinity, and thus reduce to proving the following statement in Euclidean geometry (see following page) :

Let l' be a line in the affine Euclidean plane. Let A', B', C', be three distinct point on l'. Let A, B, C be three distinct directions, different from l'. Then draw lines through A' in diections $B, C, \ldots$ and define P, Q, R as shown. Prove that P, Q, R are collinear.

We sill study various ratios : Cutting with lines in direction C, we find

$$\frac{TR}{RC'} = \frac{A'B'}{B'C'}$$

Cutting with lines of direction A, we have

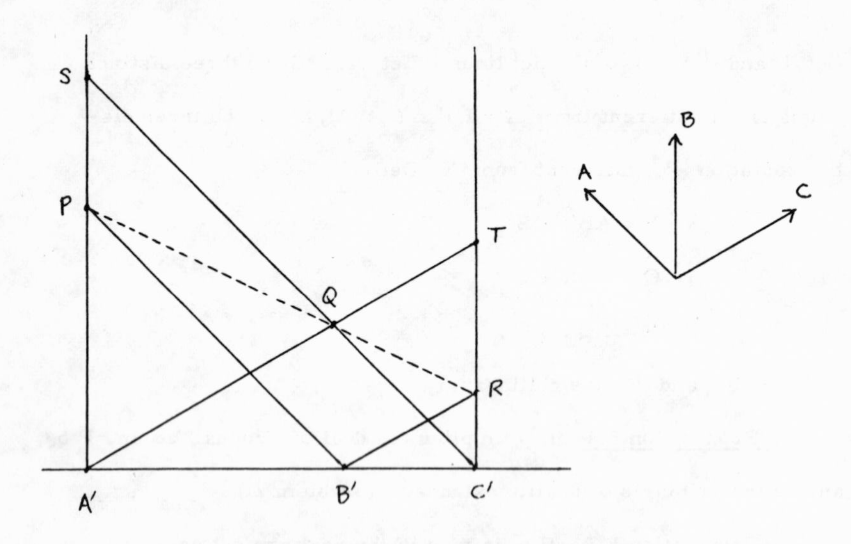

$$\frac{A'B'}{B'C'} = \frac{A'P}{PS} \ .$$

Therefore

$$\frac{TR}{RC'} = \frac{A'P}{PS} \ , \text{ or}$$

$$\frac{TR}{A'P} = \frac{RC'}{PS} = \frac{TR + RC'}{A'P + PS} = \frac{TC'}{A'S}$$

But $\Delta\, TQC' \sim \Delta\, A'QS$ (similar triangles) , so

$$\frac{TC'}{A'S} = \frac{QT}{A'Q} \ .$$

This proves that $\Delta\, TQR \sim \Delta\, A'QP$. Hence

$$\angle\, TRQ = \angle\, A'PQ \ ,$$

so PQ , QR are parallel, hence equal lines.

q. e. d.

(See problem 22 for another proof of this proposition.)

<u>Proposition 5.3</u> FT inplies P 6 (in the presence of P 1 - P 4, of course).

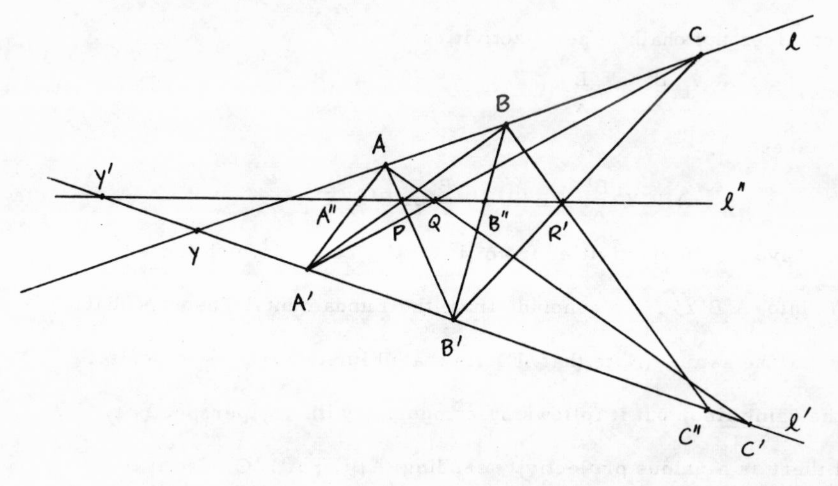

<u>Proof</u> : Let 1, 1', A, B, C, A', B', C' be as in the statement of P 6. We will assume the fundamental theorem, and will prove that

P = AB'. A'B

Q = AC'. A'C

R = BC'. B'C (not shown in diagram)

are collinear.

Draw AB', A'B , and P . Draw AC' , A'C , and Q. Let 1'' be the line PQ , and let 1'' meet AA' in A''. Then, as in Proposition 4.9 , we can construct a projectivity sending ABC to A'B'C' , as follows :

$$1 \overset{A'}{\overline{\wedge}} 1'' \overset{A}{\overline{\wedge}} 1' .$$

Let Y = 1.1', and let Y' = 1'.1'' . Then these two perspectivities act on points as follows :

$$ABCY \overset{A'}{\underset{\wedge}{=}} A''PQY' \overset{A}{\underset{\wedge}{=}} A'B'C'Y' .$$

Now let $B'C$ meet l'' in R', and let BR' meet l' in C''.

We consider the chain of perspectivities

$$l \overset{B'}{\underset{\wedge}{=}} l'' \overset{B}{\underset{\wedge}{=}} l' .$$

This takes

$$ABCY \overset{B'}{\underset{\wedge}{=}} PB''R'Y' \overset{B}{\underset{\wedge}{=}} A'B'C''Y' .$$

So we have two projectivities from l to l', each of which takes ABY into $A'B'Y'$. We conclude from the Fundamental Theorem that they are the same. (Note that FT is stated for two triples of points on the <u>same line</u>, but it follows by composing with any perspectivity that there is a unique projectivity sending $ABC \underset{\wedge}{-} A'B'C'$ also if they lie on different lines.)

Therefore the images of C must be the same under both projectivities, i.e. $C' = C''$. Therefore $R' = R$, so P, Q, R are collinear.

q.e.d.

Now we come to the proof of the Fundamental Theorem from P 1 - P 6. We must prove a number of subisidiary results first.

<u>Lemma 5.4</u> Let $l \overset{O}{\underset{\wedge}{=}} m \overset{P}{\underset{\wedge}{=}} n$, with $l \neq n$, and suppose either

a) l, m, n are concurrent, or

b) O, P and $l.n$ are collinear.

Then l is perspective to n, i.e. there is a point Q, such that the perspectivity $l \overset{Q}{=} n$ gives the same map as the projectivity $l \underset{\wedge}{-} n$ above.

<u>Proof</u>: (Problems 23, 24, and 25).

<u>Lemma 5.5</u> Let $l \overset{O}{\underset{\wedge}{=}} m \overset{P}{\underset{\wedge}{=}} n$, with $l \neq n$, and suppose that neither a) nor b) of the previous lemma holds. Then there is a line m', and points $O' \in n$, and $P' \in l$, such that

$$l \overset{O'}{\underset{\wedge}{=}} m' \overset{P'}{\underset{\wedge}{=}} n$$

gives the same projectivity from l to n.

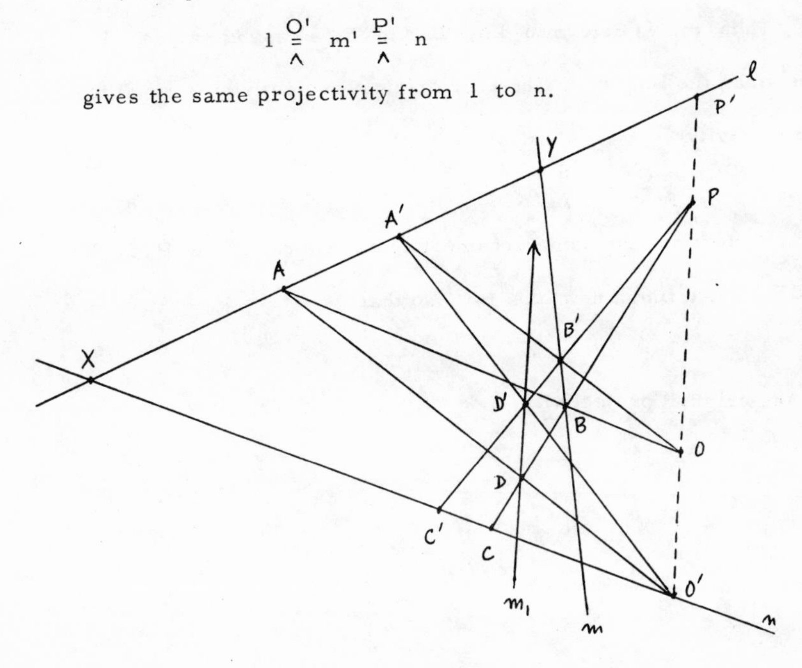

<u>Proof</u>: Let l, m, n, O, P be given. Let A, A' be two points on l, and let

$$AA' \overset{O}{\underset{\wedge}{=}} BB' \overset{P}{\underset{\wedge}{=}} CC' .$$

Let OP meet n in O'. Since we assumed $O, P, l.n = X$ are not collinear, $O' \neq X$, so $O' \notin l$. Draw $O'A$, $O'A'$, and let them meet PC, PC' in D, D', respectively.

Now correspondig sides of the triangles ABD and A'B'D' meet in O, P, O', respectively, which are collinear, hence by P 5*, the lines joining corresponding vertices are concurrent. Thus m_1, the line joining D, D', passes through the point $Y = 1.m$.

Thus m_1 is determined by D and Y, so as A' varies, D' varies along the line m_1. Thus our original projectivity is equal to the projectivity

$$1 \overset{O'}{\underset{\wedge}{=}} m_1 \overset{P}{\underset{\wedge}{=}} n.$$

Performing the same argument again, we can move P to $P' = OP.1$, and find a new line m', so that

$$1 \overset{O'}{\underset{\wedge}{=}} m' \overset{P'}{\underset{\wedge}{=}} n$$

gives the original projectivity.

Lemma 5.6 . Let l and l' be two distinct lines. Then any projectivity $l \,\overline{\wedge}\, l'$ can be expressed as the composition of two perspectivities.

Proof : A projectivity was defined as a composition of an arbitrary chain of perspectivities. Thus it will be sufficient to show, by induction , that a chain of length $n > 2$ can be reduced to a chain of length $n - 1$. Looking at one end of the chain, it will be sufficient to prove that a chain of 3 perspectivities can be reduced to a composition of two perspectivities.

The argument of the previous lemma actually shows that the line m can be moved so as to avoid any given point. Thus one can see easily (details left to reader) that it is sufficient to prove the following : let

$$l \overset{P}{\underset{\wedge}{=}} m \overset{Q}{\underset{\wedge}{=}} n \overset{R}{\underset{\wedge}{=}} o$$

be a chain of three perspectivities, with $l \neq o$. Then the resulting projectivity $l \,\overline{\wedge}\, o$ can be expressed as a product of at most two perspectivities.

First , if $m = l$ or $m = n$ or $m = o$ or $n = l$ or $n = o$, we are reduced trivially to two perspectivities, using lemma 5.4a. So we may assume l, m, n, o are all distinct. Second, using lemma 5.4b and 5.5, we have either $m \underset{\wedge}{=} o$, in which case we are done, or n can be moved so that the centers of the perspectivities $m \underset{\wedge}{=} n$ and $n \underset{\wedge}{=} o$ are on o, m respectively.

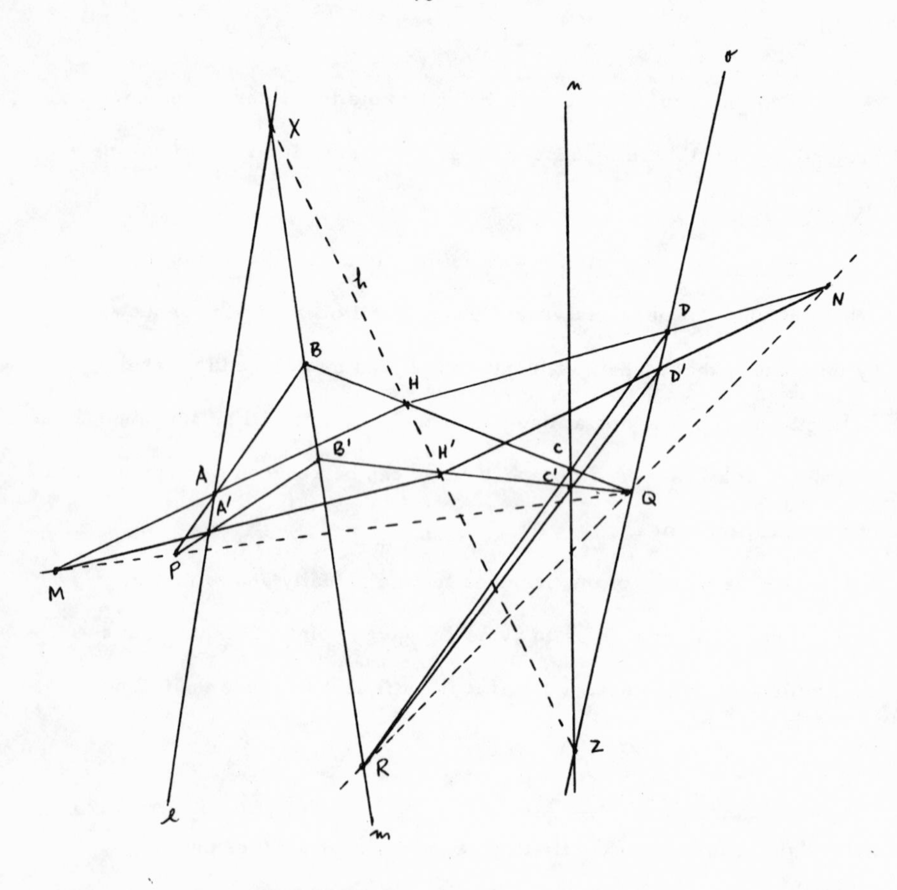

So we have

$$l \overset{P}{\underset{\wedge}{=}} m \overset{Q}{\underset{\wedge}{=}} n \overset{R}{\underset{\wedge}{=}} o$$

with l, m, n, o all distinct, $Q \in o$, and $R \in m$. Let $X = l.m$, $Z = n.o$, and draw $h = XZ$. We may assume that $X \notin o$ (indeed, we could have moved m, by lemma 5.5 to make $X \notin o$). Therefore $Q \in XZ = h$. Project $m \overset{Q}{=} h$, and let $BB' = HH'$.

Now, CDH and C'D'H' are perspective from Z. Corresponding sides meet in Q, R, hence by P 5 the remaining corresponding sides meet in a point N on QR. Thus N is determined by DH alone, and we see that as D', H' vary, the line D'H' always passes through N. In other words,

$$h \overset{N}{\underset{\wedge}{=}} o .$$

Similarly, the triangles ABH and A'B'H' are prespective from X, so using P 5 again, we find that AH and A'H' meet in a point $M \in PQ$. Hence

$$l \overset{M}{\underset{\wedge}{=}} h.$$

So we have the original projectivity represented as the composition of two perspectivities

$$l \overset{M}{\underset{\wedge}{=}} h \overset{N}{\underset{\wedge}{=}} o \quad .$$

Theorem 5.6 P 1 - P 6 imply the Fundamental Theorem.

Proof : Given a line l, and two triples of distinct points A, B, C, A', B', C', , we must show there is a unique projectivity sending ABC into A'B'C'.

Choose a line l', not passing through any of the points (I leave a few special cases to the reader), and project A', B', C', onto l'. Call them A', B', C', still. So we have reduced to the problem

$$\left.\begin{array}{l} A, B, C \text{ in } l \\ A', B', C', \text{ in } l' \end{array}\right\} \text{ all different from } l.l' .$$

It will be suffident to show that there is a unique projectivity sending $ABC \underset{\wedge}{-} A'B'C'$.

We already know one such projectivity, from Proposition 4.9. Hence it will be sufficient to show that any other such projectivity is equal to his one.

Case 1. Suppose the other projectivity is actually a perspectivity.

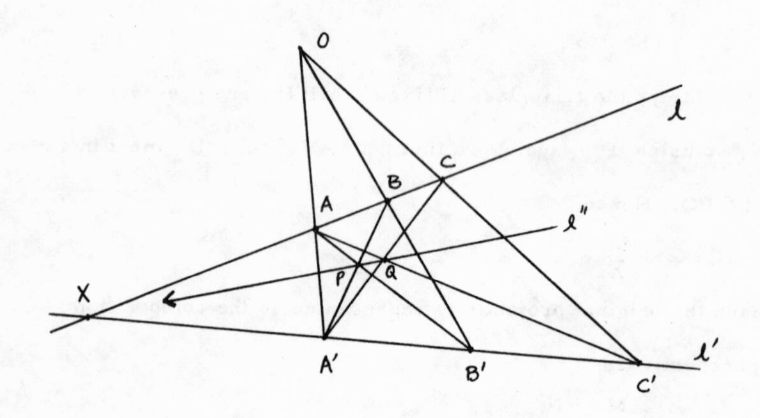

Let $l \overset{O}{\underset{\wedge}{=}} l'$ send $ABC \underset{\wedge}{=} A'B'C'$. Consider

$$P = AB' \cdot A'B$$

$$Q = AC' \cdot A'C$$

and let l'' be the line joining P and Q.

I claim that l'' passes through X. Indeed, we apply P 5 to the two triangles $AB'C'$ and $A'BC$, which are perspective from O. Their corresponding sides meet in P, Q, X respectively.

Hence l'' is already determined by P and X. This shows, that as C varies, the perspectivity

$$l \overset{Q}{\underset{\wedge}{=}} l'',$$

and the projectivity

$$1 \overset{A'}{\underset{\wedge}{=}} 1'' \overset{A}{\underset{\wedge}{=}} 1'$$

coincide.

<u>Case 2.</u> Suppose the other projectivity is not a perspectivity. Then by lemma 5.6 , it can be expressed as the composition of (exactly) two perspectivities, and by lemma 5.5, we can assume that their centers lie on 1' and 1, respectively. Thus we have the following diagram :

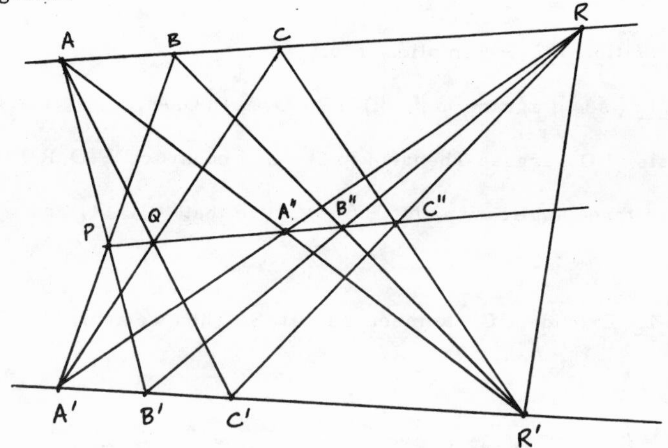

Here $1 \overset{R'}{\underset{\wedge}{=}} 1'' \overset{R}{\underset{\wedge}{=}} 1'$, and $ABC \overset{R'}{\underset{\wedge}{=}} A''B''C'' \overset{R}{\underset{\wedge}{=}} A'B'C'$. By P 6 applied to ABR and A'B'R' , the point

$$P = AB' \, . \, A'B$$

lies on 1''. Similarly by P 6 applied to ACR and A'C'R' ,

$$Q = AC' \, . \, A'C$$

lies on 1'' . Thus 1''is the line which was used in Proposition 4.9 to construct the other projectivity

$$l \overset{A'}{\underset{\wedge}{=}} l'' \overset{A}{\underset{\wedge}{=}} l'.$$

Now if $D \in l$ is an arbitrary point, define $D'' = R'D.l''$, and $D' = RD''.l'$. Then consider P 6 applied to ADR and A'D'R'. It says

$$AD' . A'D , A'' , D''$$

are collinear, i.e. $AD'.A'D \in l''$, which means that D goes into D' also by the projectivity of Proposition 4.9. Hence the two projectivities are equal.

q. e. d.

Proposition 5.7 P 6 implies P 5.

Proof : (see diagram on p. 81) Let O, A, B, C, A', B', C' satisfy the hypothesis of Desargues Theorem (p 5), and construct P, Q, R. We will make three applications of P 6 to prove that P, Q, R, are collinear.

Step 1. Extend A'C' to meet AB at S. Then we apply P 6 to the lines

$$\begin{pmatrix} O & C & C' \\ B & S & A \end{pmatrix}$$

and conclude that

$$T = OS . BC$$

$$U = OA . BC'$$

$$Q$$

are collinear. (Note to apply P 6, we should check that B, S, A are all distinct, and O, C, C', B, S, A are all different from the intersection of the two lines. But P 6 is trivial if not.)

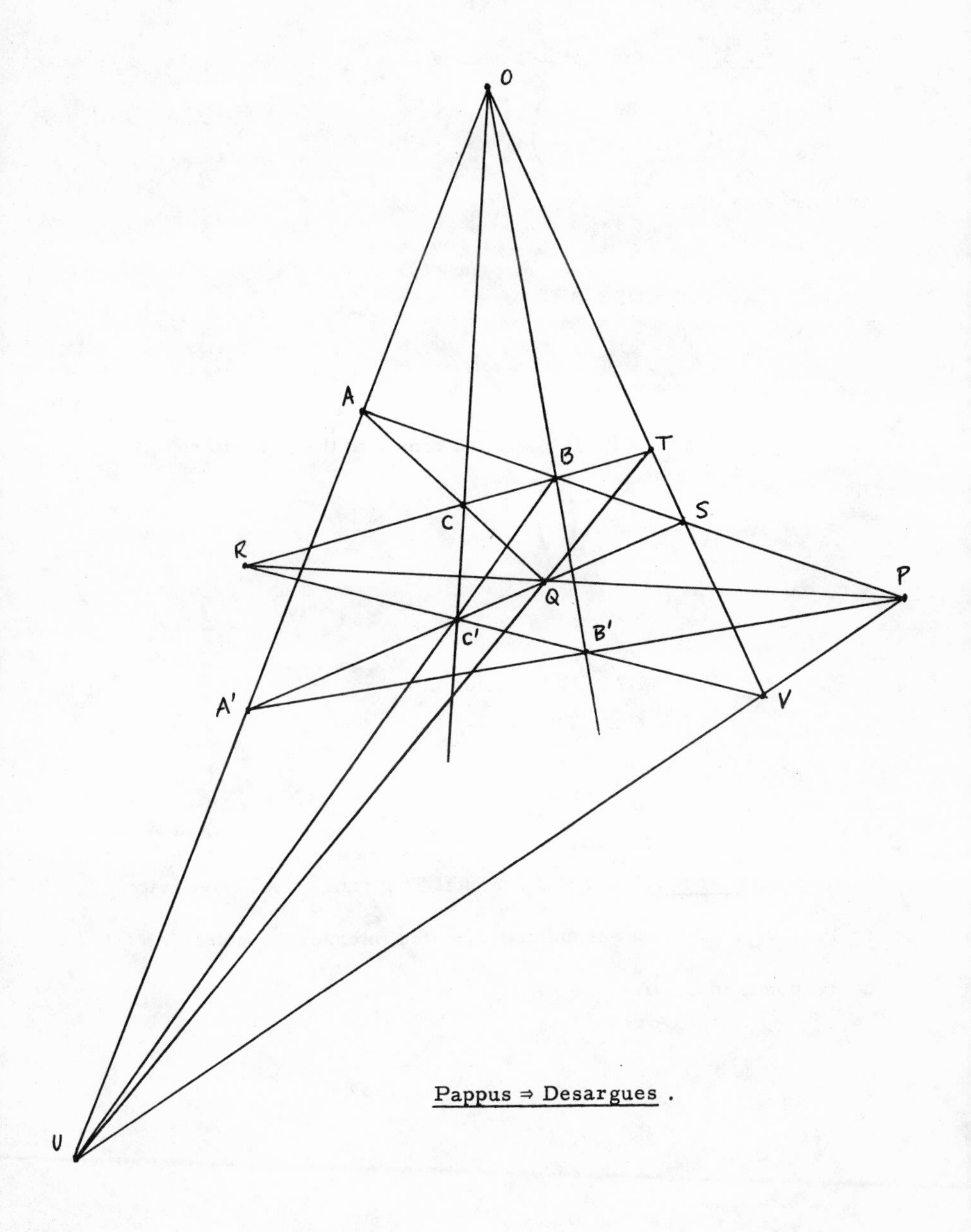

Pappus ⇒ Desargues .

Step 2. We apply P 6 a second time, to the two triples

$$\begin{pmatrix} O & B & B' \\ C' & A' & S \end{pmatrix}$$

and conclude that

U

V = OS . B'C'

P

are collinear.

Step 3. We apply P 6 a third time , to the two triples

$$\begin{pmatrix} B & C' & U \\ V & T & S \end{pmatrix}$$

and conclude that

R

P = BS . UV by Step 2.

Q = C'S . TU by Step 1.

are collinear.

q. e. d.

Corollary 5. 8 (of Fundamental Theorem) . A projectivity $l \,\overline{\wedge}\, l'$ with $l \neq l'$ is a perspectivity $\Leftrightarrow$ the intersection point $X = l . l'$ corresponds to itself.

CHAPTER 6 : PROJECTIVE PLANES OVER DIVISION RINGS

In this chapter we intorduce the notion of a division ring, which is slightly more general than a field, and the projective plane over a division ring. This will give us many examples of projective planes, besides the ones we know already. Then we will discuss various properties of the projective plane corresponding to properties of the field. We will also study the group of automorphisms of these projective planes.

Definition : A division ring (or skew field , or sfield, or non-commutative field) , is a set F, together with two operations $+$ and $\cdot$, such that

1) F is an abelian group under $+$

2) The non-zero elements of F form a group under $\cdot$, (not necessarily commutative) , and

3) Multiplication is distributive over addition , on both sides, i. e. for all $a, b, c \in F$, we have

$$a(b + c) = ab + ac$$

$$(b + c)a = ba + ca .$$

Comparing with the definition of a field on p. 39 , we see that a division ring is a field $\Leftrightarrow$ the commutative law for multiplication holds.

Example : (to show that there are some division rings which are not fields) . We define the division ring of quaternions as follows. Let e, i, j, k be four symbols. Define

$$F = \{ ae + bi + cj + dk \mid a, b, c, d \in \mathbb{R} \} .$$

We make F into a division ring by adding place by place :

$$(ae + bi + cj + dk) + (a'e + b'i + c'j + d'k) =$$

$$= (a + a')e + (b + b')i + (c + c')j + (d + d')k .$$

We define multiplication by

a) using the distributive laws

b) decreeing that the ral number commute with everything else,

and

c) multiplying e, i, j, k according to the following table :

$$e^2 = e$$

$$i^2 = j^2 = k^2 = -e$$

$$e.i = i.e = i$$

$$e.j = j.e = j$$

$$e.k = k.e = k$$

$$i.j = k \qquad j.i = -k$$

$$j.k = i \qquad k.j = -i$$

$$k.i = j \qquad i.k = -j .$$

Then one can check (rather laboriously) that F is a division ring. And of course it is not a field, because multiplication is not commutative. e.g. $ij \neq ji$.

Definition. An automorphism of a division ring is a 1 - 1 mapping $\sigma : F \longrightarrow F$ of F onto F (which we will write $a \longrightarrow a^\sigma$) such that

$$(a + b)^{\sigma} = a^{\sigma} + b^{\sigma}$$

$$(ab)^{\sigma} = a^{\sigma} b^{\sigma} .$$

Definition . Let F be a division ring. The characteristic of F is the smallest integer $p \geq 2$ such that

$$\underbrace{1 + 1 + \ldots + 1}_{p \text{ times}} = o,$$

or, if there is no such integer, the characteristic of F is defined to be o.

Proposition : The characteristic p of a division ring F is always a prime number.

Proof : Suppose $p = m \cdot n$. $m, n > 1$. Then

$$\underbrace{(1 + 1 + \ldots + 1)}_{m \text{ times}} \cdot \underbrace{(1 + 1 + \ldots + 1)}_{n \text{ times}} = o$$

Hence one of them is o, which contradicts the choice of p.

Example : For any prime number p, there is a field F_p with p elements , and having characteristic p. Indeed, let F_p be the set of p symbols $F = \{o, 1, 2, \ldots, p-1\}$. Define addition and multiplication in F by treating the symbols as integers, and then reducing modulo p. (For example $2 \cdot (p-1) = 2p-2 \equiv p-2 \pmod{p}$.) Then F is a field, as one can check easily, and has characteristic p.

Definition : Let F be a division ring, and let $F_0 \subseteq F$ be the set of $a \in F$ such that $ab = ba$ for all $b \in F$. Then F_o is a field, and it is called the center of F.

To see that F_o is a field, we must check that it is closed under addition, multiplication, taking of inverses, and that the commutative law of multiplication holds. These are all easy. For example, say $a, b \in F_o$. Then for any $c \in F$,

$$(a + b)c = ac + bc = ca + cb = c(a + b),$$

so $a + b \in F_o$.

<u>Example</u>: The center of the division ring of quaternions is the set of quaternions of the form

$$a \cdot e + o \cdot i + o \cdot j + o \cdot k ,$$

for $a \in \mathbb{R}$. Hence $F_o \cong \mathbb{R}$.

Now we can define the projective plane over a division ring, mimicking the analytic definition of the real projective plane (p. 9).

<u>Definition</u>. Let F be a division ring. We define the <u>projective plane over F</u>, written $\mathbb{P}_F^2$, as follows. A <u>point</u> of the projective plane is an equivalence class of triples

$$P = (x_1, x_2, x_3)$$

where $x_1, x_2, x_3 \in F$, are not all zero, and where two triples are equivalent,

$$(x_1, x_2, x_3) \sim (x_1', x_2', x_3')$$

if and only if there is an element $\lambda \in F$, $\lambda \neq o$, such that

$$x_i' = x_i \lambda \quad \text{for } i = 1, 2, 3.$$

(Note that we multiply by λ on the <u>right</u>. It is important to keep this in mind, since the multiplication may not be commutative.)

A line in $\mathbb{P}_F^2$ is the set of all points satisfying a linear equation of the form

$$c_1x_1 + c_2x_2 + c_3x_3 = o$$

where $c_1, c_2, c_3 \in F$, and are not all zero. Note that we multiply here on the left, so that this equation actually defines a set of equivalence classes of triples.

Now one can check that the axioms P 1, P 2, P 3, P 4 are satisfied, and so $\mathbb{P}_F^2$ is a projective plane.

Examples . 1. If $F = F_2$ is the field of two elements $\{o, 1\}$, then $\mathbb{P}_F^2$ is the projective plane of seven points.

2. More generally, if $F = F_p$ for any prime number p, then $\mathbb{P}_F^2$ is a projective plane with $p^2 + p + 1$ points. Indeed, any line has $p + 1$ points, so this follows from problem 5.

3. If $F = \mathbb{R}$ we get back the real projective plane.

Theorem 6.1 The plane $\mathbb{P}_F^2$ over a division ring always satisfies Desargues ' axiom P 5.

Proof : One defines projective 3-space $\mathbb{P}_F^3$ by taking points to be equivalence classes $(x_1, x_2, x_3, x_4) = x_i \in F$, not all zero, and where this is equivalent ot $(x_1\lambda, x_2\lambda, x_3\lambda, x_4\lambda)$. Planes are defined by (left) linear equations, and lines as intersections of distinct planes.

Then $\mathbb{P}_F^2$ is embedded as the plane $x_4 = o$ in this projective 3-space , and so P 5 holds there by an earlier result. (theorem 2.1) .

Now we will study the group <u>$\mathrm{Aut}(\mathbb{P}_F^2)$</u> of automorphisms of our projective plane.

<u>Definition :</u> A matrix $A = (a_{ij})$ of elements of F is <u>invertible</u> if there is a matrix A^{-1}, such that $AA^{-1} = A^{-1}A = I$, the identity matrix. (Note that in general determinants do not make sense over a division ring. However, if we are working over a field F, these are just the matrices with determinant $\neq 0$.)

<u>Proposition 6.2</u>. Let $A = (a_{ij})$ be an invertible 3×3 matrix of elements of F. Then the equations

$$x_i' = \sum_{j=1}^{3} a_{ij} x_j \qquad i = 1, 2, 3$$

define an automorphism T_A of $\mathbb{P}_F^2$.

<u>Proof</u>. Analogous to proof of Proposition 3.7 q.v.

<u>Proposition 6.3</u>. Let A, A' be two invertible matrices. Then T_A and $T_{A'}$ have the same effect on the four points P_1, $(1,0,0)$, $P_2 = (0,1,0)$, $P_3 = (0,0,1)$, $Q = (1,1,1)$ $\Leftrightarrow$ there is a $\lambda \in F$, $\lambda \neq 0$, such that $A' = A\lambda$.

<u>Proof :</u> Analogous to Propositions 3.8 q.v.

<u>Proposition 6.4.</u> Let $\lambda \in F$, $\lambda \neq 0$, and consider the matrix λI, Then $T_{\lambda I}$ is the identity transformation of $\mathbb{P}_F^2$ $\Leftrightarrow$ λ is in the center of F. Otherwise, $T_{\lambda I}$ is the automorphism given by

$$(x_1, x_2, x_3) \longrightarrow (x_1^\sigma, x_2^\sigma, x_3^\sigma) ,$$

where σ is the automorphism of F given by

$$x \longrightarrow \lambda x \lambda^{-1} .$$

(Such an automorphism is called an <u>inner automorphism of F.</u>)

<u>Proof :</u> In general, $T_{\lambda I}$ takes (x_1, x_2, x_3) to the point $(\lambda x_1, \lambda x_2, \lambda x_3)$. This latter point also has homogeneous coordinates $(\lambda x_1 \lambda^{-1}, \lambda x_2 \lambda^{-1}, \lambda x_3 \lambda^{-1})$, which proves the second assertion. But σ is the identity automorphism of $F \Leftrightarrow \lambda x = x\lambda$ for all x, i.e. λ is in the center of F .

<u>Corollary 6.5.</u> Let A and A' be invertible matrices, Then $T_A = T_{A'} \Leftrightarrow \exists\, \lambda \in$ center of F, $\lambda \neq 0$, such that $A' = A\lambda$.

<u>Proof :</u> $\Leftarrow$ is clear. Conversely, if $T_A = T_{A'}$, then by Proposition 6.3 , $A' = A\lambda = A \cdot (\lambda I)$. So $T_{A'} = T_A \cdot T_{\lambda I}$, so $T_{\lambda I}$ is identity so $\lambda \in$ center of F.

<u>Definition</u>. We denote by <u>PGL(2, F)</u> the group of automorphism of $\mathbb{P}^2_F$ of the form T_A for some invertible matrix A. (Thus PGL(2, F) is the quotient of the group GL(3, F) of invertible matrices, by multiplication by scalars in the center of F.)

<u>Proposition 6.6.</u> Let A, B, C, D and A', B', C', D' be two quadruples of points, no 3 collinear. Then there is an element $T \in$ PGL(2, F) such that $T(A) = A'$, $T(B) = B'$, $T(C) = C'$, $T(D) = D'$.

<u>Proof :</u> Analogous to Theorem 3.9 q.v.

Note that in general the transformation T is not unique. However, if F is commutative, it will be unique, by Proposition 6.3 and Corollary 6.5, Since F is its own center.

Proposition 6.7. Let φ be any automorphism of $\mathbb{P}_F^2$ which leaves fixed the four points P_1, P_2, P_3, Q mentioned above. Then there is an automorphism $\sigma \in \text{Aut } F$, such that

$$\varphi(x_1, x_2, x_3) = (x_1^\sigma, x_2^\sigma, x_3^\sigma) .$$

Proof : Analogous to Proposition 3.11 q.v. (Except that instead of using Euclidean methods in the proof, one must show by analytic geometry over F that the constructions for $a + b, ab$ work.)

Proposition 6.8. The mapping $\text{Aut } F \longrightarrow \text{Aut } \mathbb{P}_F^2$ given by $\sigma \longrightarrow$ the map φ described in the previous Proposition is an isomorphism of Aut F onto the subgroup H of $\text{Aut } \mathbb{P}_F^2$ consisting of those automorphisms which leave P_1, P_2, P_3, Q fixed.

Proof : It is onto by the previous Proposition. To see it is 1 - 1, apply σ and $\sigma' \in \text{Aut } F$ to $(x, 1, o)$. Then $(x^\sigma, 1, o)$ is the same point as $(x^{\sigma'}, 1, o)$, so $x^\sigma = x^{\sigma'}$, and $\sigma = \sigma'$. Clearly it perserves the group law.

We can sum up all our information about $\text{Aut } \mathbb{P}_F^2$ in the diagram on the top of the next page. The two subgroups PGL(2, F) and H generate $\text{Aut } \mathbb{P}_F^2$, i.e. every element of the whole group can be expressed as a product of elements in the two subgroups. (This follows from Propositions 6.6 and 6.7). The intersection K of the two subgroups is isomorphic to the group of inner automorphisms of F. (by Propositions 6.3 and 6.4).

Now we will see when the axioms P 6 and P 7 hold in a

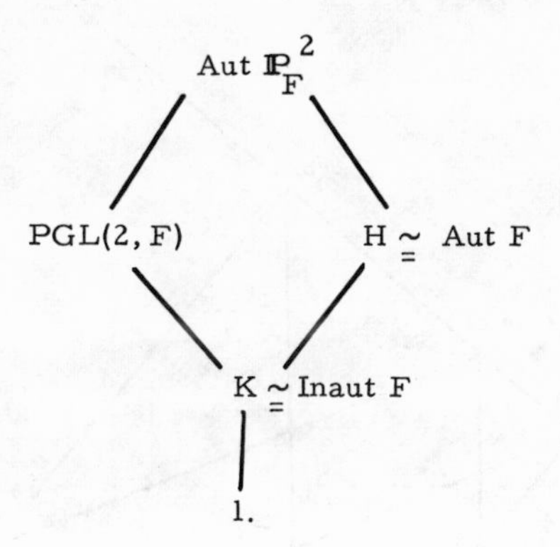

projective plane $\mathbb{P}_F^2$.

Theorem 6. 9. Pappus' axiom , P 6, holds in the projective plane $\mathbb{P}_F^2$ over a division ring $F \Leftrightarrow F$ is commutative.

Proof : First let us suppose that P 6 holds. We take $x_3 = o$ to be line at infinity, and represent an element $a \in F$ as the point (a, o) on the x-axis . If (a, o) , (b, o) are two points, we can construct the product of a and b with the diagram of page 43 . However, this time we are working over the division ring F, not over the real numbers, so we must verify analytically that the construction works.

By inspection , one finds that the equation of the line joining $(1, 1)$ and (b, o) is

$$x + (b-1)y = b.$$

Hence the equation of the line parallel to this one, through (a, a) is

$$x + (b-1)y = ba,$$

so that the point we have constructed is (ba, o) .

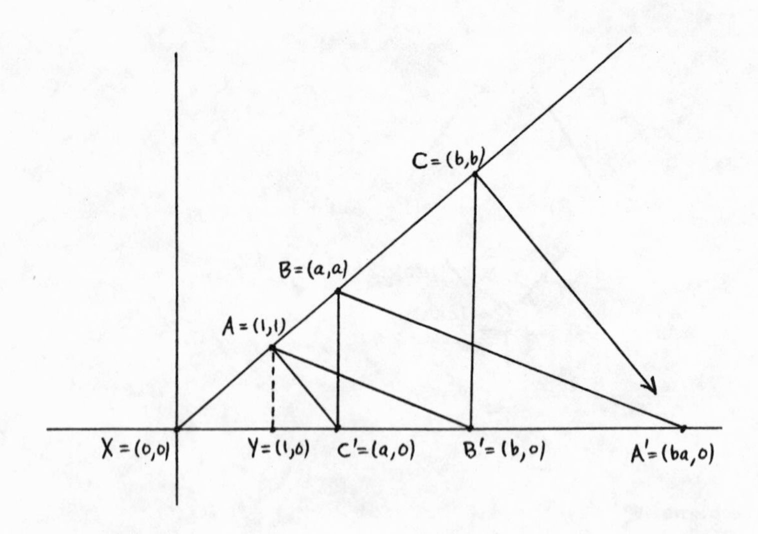

To get the product in the other order, we reverse the process, by drawing the line through (1, 1) and (a, o) , and the line parallel to this through (b, b) . Now the affine version of P6 implies that we get the same point. Hence ab = ba, and F is commutative.

Before proving the converse, we give a lemma.

<u>Lemma 6.10.</u> Let l, A, B, C , and l', A', B', C' be two sets, each consisting of a line, and three non-collinear points, not on the line, in $\mathbb{P}_F^2$. Then there is an automorphism φ of $\mathbb{P}_F^2$ such that $\varphi(l) = l'$, and $\varphi(A) = A'$, $\varphi(B) = B'$, $\varphi(C) = C'$.

<u>Proof :</u> Let X = l. AC and Y = l. BC , and define similarly X' = l'. A'C' , Y' = l'. B'C' , Then A, B, X, Y are four points, no three collinear, and similarly for A', B', X', Y' , so by Proposition 6. 6 there is an automorphism φ of $\mathbb{P}_F^2$ sending A, B, X, Y into

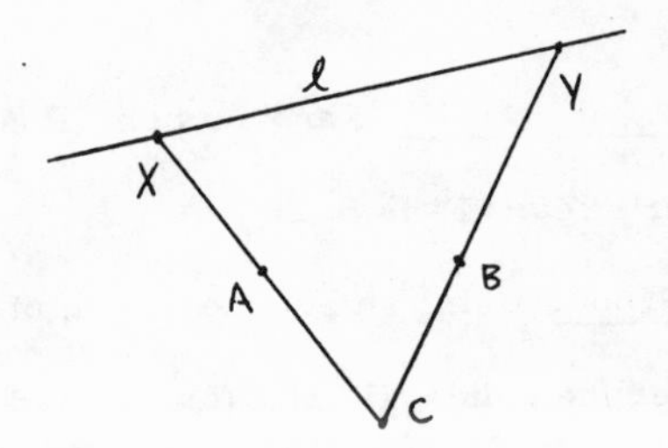

A', B', X', Y'. Then clearly φ sends l into l ' and C into C'.

Proof of Theorem 6.9 continued. Now assume F is commutative, and let us prove P 6.

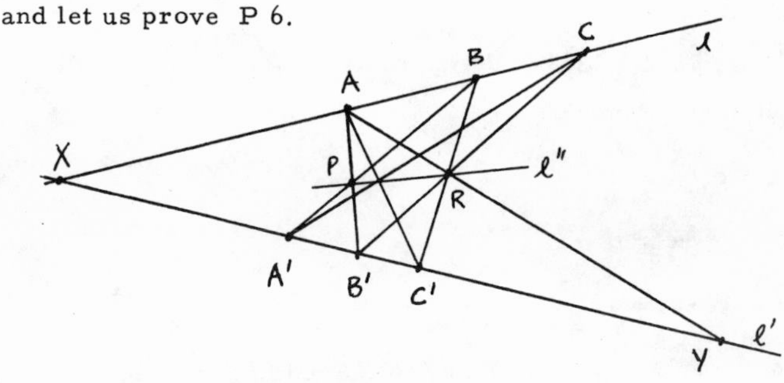

With the usual notation, let P = AB'.A'B, R = BC'.B'C, and let l'' be the line PR. We may assume that X = l.l' does not lie on l''. (If it did, take a different pair P, Q or Q, R. If all these three pairs lie on lines through X, then P, Q, R are already collinear, and there is nothing to prove.) Let Y = AR.l'. Then Y is not on l'', and A, X, Y are non-collinear. Hence by the lemma, we can find an automorphism φ of $\mathbb{P}_F^2$ taking l'' to the line $x_3 = o$, and taking A, X, Y to the points (1, 1), (o, o), (1, o), respectively.

Then we have the situation of the diagram on page 92 again, where we wish to prove AC' $\|$ A'C. But this follows from the commutativity of F.

<u>Theorem 6.11.</u> Fano's axiom P 7 holds in $\mathbb{P}_F^2 \Leftrightarrow$ the characteristic of $\mathbb{F}$ is $\neq 2$.

<u>Proof :</u> Using an automorphism of $\mathbb{P}_F^2$, we reduce to the question of whether the points (1, 1ρ), (1, o, 1), and (o, 1, 1) are collinear, as in the proof of Proposition 4.3. Since F may not be commutative, we will not use matrices, but will give a direct proof. Suppose they are collinear. Then they all satisfy an equation

$$c_1x_1 + c_2x_2 + c_3x_3 = o,$$

with the c_i not all zero. Hence

$$\begin{aligned} c_1 + c_2 &= o \\ c_1 + c_3 &= o \\ c_2 + c_3 &= o \end{aligned}$$

Thus $c_1 = -c_2$, $c_1 = -c_3$, $c_2 = c_3$, so $2c_2 = o$. So either $c_2 = o$, in which case $c_3 = o$, $c_1 = o$ ✱, or $2 = o$, in which case the characteristic of F is 2.

As a dessert, we are now in a position to show that among the axioms P 5, P 6, P 7, the only implication is P 6 ⇒ P 5 (Proposition 5.7). We prove this by giving examples of projective planes which have every possible combination of axioms holding or not.

<u>Explanations</u>

1) The projective plane of seven points has P 5, P 6, not P 7.

2) The real projective plane $\mathbb{P}_{\mathbb{R}}^2$ has P 5, P 6, P 7.

3) The free projective plane on 4 points has not P 5, not P 6, P 7.

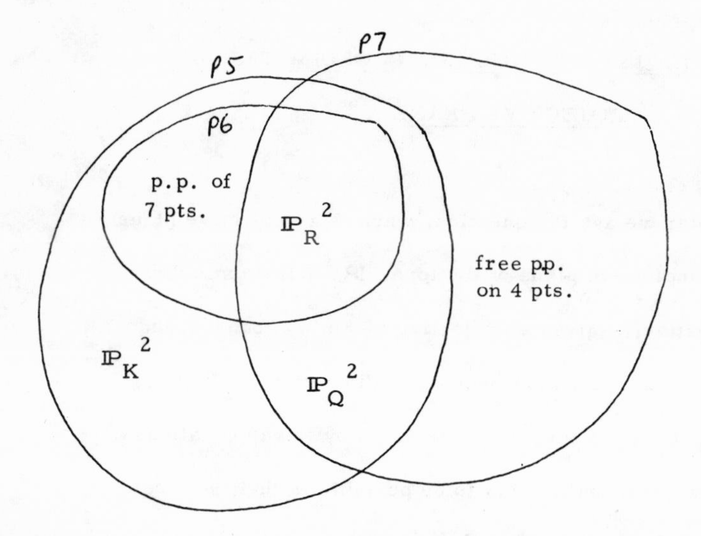

free p.p on (p.p. of 7 points) $\cup$ {one point} .

4) Let Q be the division ring of quaternions . Then $\mathbb{P}_Q^2$ has P 5, not P 6, P 7 , since char. Q = o.

5) Let K be a non-commutative division ring of char.. 2. (one can obtain one of these as follows : Let $k = \{o, 1\}$,let $k[t]$ be the ring of polynomials in t with coefficients in k, let α be the endomorphism of $k[t]$ defined by $t \longrightarrow t^2$, let $A = \{ \sum_{i=1}^{n} p_i(t) X^i \}$, where X is an indeterminate, and make A into a ring by defining $Xp(t) = \alpha(p(t))X$. Then one can show that A can be embedded in a division ring K, which is necessarily non-commutative). Then $\mathbb{P}_K^2$ has P 5, not P 6, not P 7.

6) Let π_o be a projective plane of 7 points, plus one extra point with no lines. Then the free projective plane on π_o satisfies not P 5, not P 6, not P 7.

CHAPTER 7. INTRODUCTION OF COORDINATES IN A PROJECTIVE PLANE.

In this chapter we ask the question, when is a projective plane π isomorphic to a projective plane of the form $\mathbb{P}_F^2$, for some division ring F ? Or alternatively, given a projective plane π, can we find a division ring F, and assign homogeneous coordinates (x_1, x_2, x_3), $x_i \in F$, to points of π, such that lines are given by linear equations ?

A necessary condition for this to be possible is that π should satisfy Desargues' axiom, P 5, since we have seen that $\mathbb{P}_F^2$ always satisfies P 5 (Theorem 6.1). And in fact we will see that Desargues' axiom is sufficient.

We will begin with a simpler problem, namely the introduction of coordinates in an affine plane A. A naive approach to this problem would be the following : Choose three non-collinear points in A, and call them $(1, o)$, (o, o), $(o, 1)$. Let l be the line through (o, o) and $(1, o)$. Now take F to be the set of points on l, and define addition and multiplication in F be the geometrical construction given in the proof of Proposition 3.11 (pp. 40-45). Then one would have to verify that F was a division ring, i. e. prove that addition was commutative and associative, that multiplication was associative and distributive, etc. The proofs would involve some rather messy diagrams. Then finally one would coordinatize the plane using these coordinates on l,

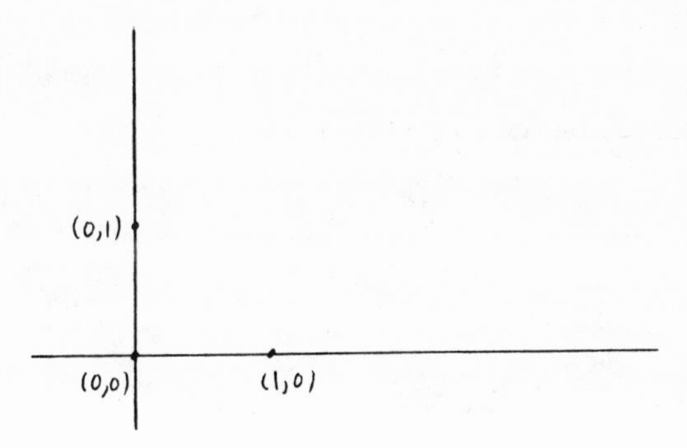

and prove that lines were given by linear equations. In fact, this is the approach which is used in Seidenberg's book, "Lectures in Projective Geometry", chapter 3.

However, we will use a slightly more sophisticated method, on the principle that if one uses more high-powered techniques, there will be less work to be done. Hence we will first address ourselves to a study of certain automorphisms of an affine plane.

Definition : Let A be an affine plane. A dilatation is an automorphism φ of A, such that for any two distinct points P, Q, $PQ \parallel P'Q'$, where $\varphi(P) = P'$, $\varphi(Q) = Q'$. In other words, φ takes lines into parallel lines. Or, if we think of A as contained in a projective plane $\pi = A \cup l_\infty$, then φ is an automorphism of π, which leaves the line at infinity, l_∞, pointwise fixed.

Examples : In the real affine plane $\mathbb{A}_{\mathbb{R}}^2 = \{(x,y) \mid x, y \in \mathbb{R}\}$, a stretching in the ratio k, given by equations

$$\begin{cases} x' = kx \\ y' = ky \end{cases}$$

is a dilatation. Indeed, let O be the point (o, o)

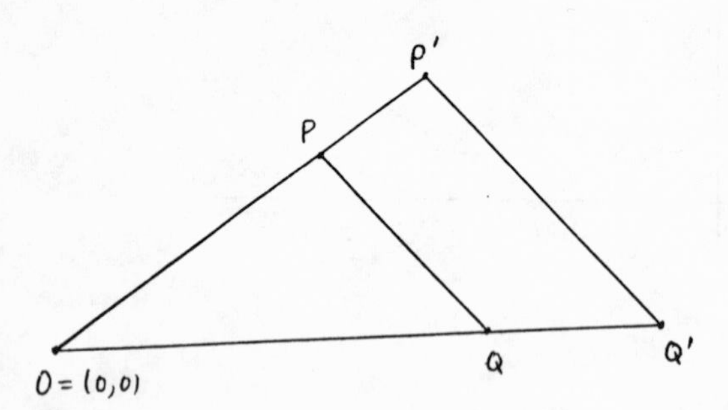

Then φ stretches points away from O k-times, and if P, Q are any two points, clearly $PQ \parallel P'Q'$, by similar triangles.

Another example of a dilatation of $\mathbb{A}^2_{\mathbb{R}}$ is given by a translation

$$\begin{cases} x' = x + a \\ y' = y + b \end{cases}$$

In this case, any point P is translated by the vector from O to (a, b), so $PQ \parallel P'Q'$ again, for any P, Q.

Without asking for the moment whether there are any non-trivial dilatations in a given affine plane A, let us study some of their properties.

<u>Proposition 7.1.</u> Let A be an affine plane. Then the set of dilatations, <u>Dil (A)</u> forms a subgroup of the group of all automorphisms of A, Aut A.

<u>Proof :</u> Indeed, we must see that the product of two dilatations is a dilatation, and that the inverse of a dilatation is a dilatation. This follows immediately from the fact that parallelism is an equivalence relation.

<u>Proposition 7.2.</u> A dilatation which leaves two distinct points fixed is the identity.

<u>Proof :</u> Let φ be a dilatation, let P, Q be fixed, and let R be any point not on PQ. Let $\varphi(R) = R'$. Then we have

$$PR \parallel PR'$$

and

$$QR \parallel QR'$$

since φ is a dilatation. Hence $R' \in PR$ and $R' \in OR$. But $PR \nparallel QR$ since $R \notin PQ$. Hence $PR \cdot QR = \{R\}$, and so $R = R'$, i.e. R is also fixed. But R was an arbitrary point not on PQ. Applying the same argument to P and R, we see that every point of PQ is also fixed, so φ is the identity.

<u>Corollary 7.3.</u> A dilatation is determined by the images of two points, i.e. any two dilatations φ, ψ , which behave the same way on two distinct points P, Q, are equal.

<u>Proof :</u> Indeed, $\psi^{-1}\varphi$ leaves P, Q fixed, so is the identity.

So we see that a dilatation different from the identity can have at most one fixed point. We have a special name for those dilatations with no fixed points :

Definition . A translation is a dilatation with no fixed points, or the identity .

Proposition 7.4. If φ is a translation, different from the identity, then for any two points P, Q , we have $PP' \parallel QQ'$, where $\varphi(P) = P'$, $\varphi(Q) = Q'$.

Proof : Suppose $PP' \nparallel QQ'$. Then these two lines intersect in a point O. But the fact that φ is a dilatation implies that φ sends the line PP' into itself, and φ sends QQ' into itself. (Forexample , let $R \in PP'$. Then $PR \parallel P'R'$, but $PR = PP'$, so $R \in PP'$.) Hence $\varphi(O) = O$, so O is a fixed point ✱.

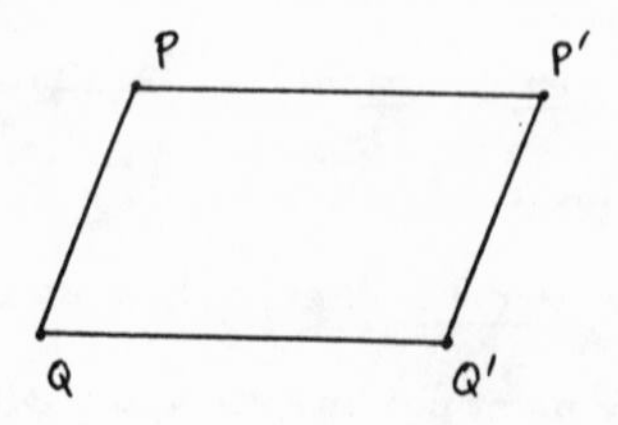

Proposition 7.5. The translations of A form a subgroup Tran (A) of the group of dilatations of A. Furthermore, Tran(A) is a normal subgroup of Dil(A) , i.e. for any $\tau \in$ Tran(A) , and $\sigma \in$ Dil(A) ,

$$\sigma\tau\sigma^{-1} \in \mathrm{Tran(A)}.$$

Proof: First we must check that the product of two translations is a translation, and the inverse of a translation is a translation. Let τ_1, τ_2 be translations, then $\tau_1\tau_2$ is a dilatation. Suppose it has a fixed point P. Then $\tau_2(P) = P'$, $\tau_1(P') = P$. If Q is any point not on PP' , then let $Q' = \tau_2(Q)$.

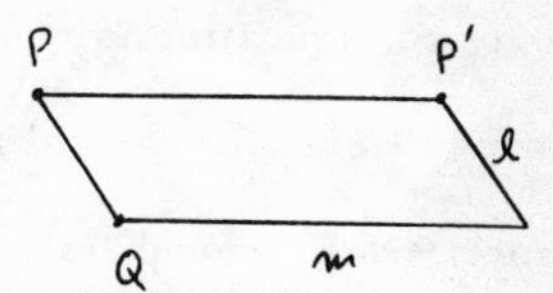

We have by the previous Proposition

$$PQ \parallel P'Q' \text{ and } PP' \parallel QQ'.$$

Hence Q' is determined as the intersection of the line $l \parallel PQ$, through P' , and the line $m \parallel PP'$ through Q.

For a similar reason, $\tau_1(Q') = Q$. Hence Q is also fixed. Applying the same reasoning to Q, we find every point is fixed, so $\tau_1\tau_2 = \text{id}$. Hence $\tau_1\tau_2$ is a translation. Clearly the inverse of a translation is a translation, so the translations form a subgroup of Dil(A).

Now let $\tau \in \text{Tran}(A)$. $\sigma \in \text{Dil}(A)$. Then $\sigma\tau\sigma^{-1}$ is certainly a dilatation. If it has no fixed points, it is a translation, ok. If it has a fixed point P , then $\sigma\tau\sigma^{-1}(P) = P$ implies $\tau\sigma^{-1}(P) = \sigma^{-1}(P)$, so τ has a fixed point . Hence $\tau = \text{id}$, and $\sigma\tau\sigma^{-1} = \text{id}$, ok .

Definition . In general, if G is a group, and H is a subgroup of G, we say H is a normal subgroup of G if $\forall\, h \in H$ and $\forall\, g \in G$,

$$ghg^{-1} \in H .$$

For example, in an abelian group, every subgroup is normal.

Now we come to the question of existence of translations and dilatations, and for this we will need Desargues' axiom . In fact, we will find that these two existence problems are equivalent to two affine forms of Desargues' axiom . This is one of those cases where an axiom about some configuration is equivalent to a property of the geometry of the space. Here Desargues' axiom is equivalent to saying that our geometry has "enough" automorphisms in a sense which will become clear form the theorems.

A 5 a (Small Desargues' axiom.) Let l, m, n be three parallel lines (distinct). Let $A, A' \in l$, $B, B' \in m$, $C, C' \in n$, all distinct points. Assume $AB \parallel A'B'$, and $AC \parallel A'C'$. Then $BC \parallel B'C'$.

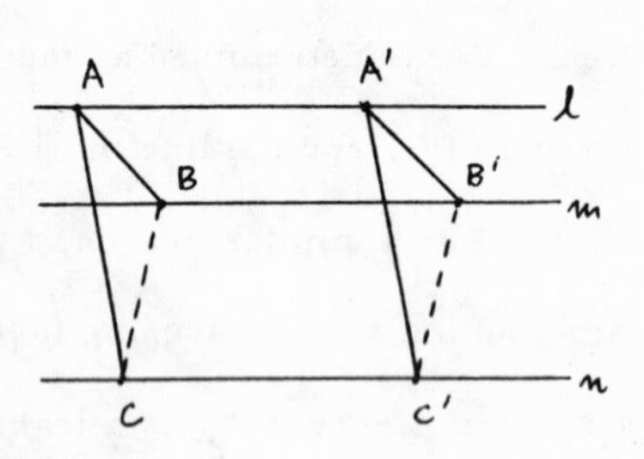

Note that if our affine plane A is contained in a projective plane π, then A5a follows form P 5 in π. Indeed, l, m, n meet in a point O on the line at infinity l_∞. Our hypotheses state that

$$P = AB \, . \, A'B' \in l_\infty$$

$$Q = AC \, . \, A'C' \in l_\infty \, .$$

So P 5 says that

$$R = BC \, . \, B'C' \in l_\infty \, ,$$

i.e. $BC \parallel B'C'$.

Theorem 7.6. Let A be an affine plane. Then the following two statements are equivalent :

i) The axiom A5a holds in A

ii) Given any two points $P, P' \in A$, there exists a unique translation τ such that $\tau(P) = P'$.

Proof : (i) ⇒ (ii) We assume A5a . If $P = P'$, then the identity is a translation taking P to P' , and it is the only one, so there is nothing to prove. So suppose $P \neq P'$.

Now we will set out to construct a translation τ sending P to P' .

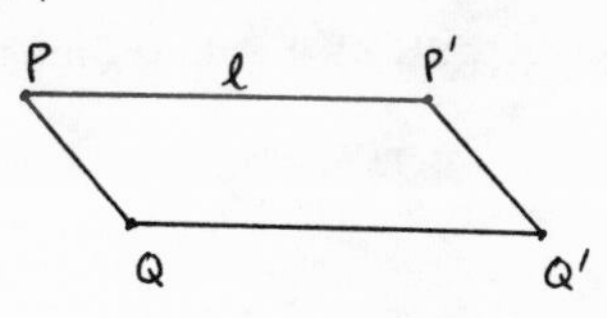

Step 1 We define a transformation $\tau_{PP'}$ of $A - l$, where l is the line PP' , as follows : for $Q \notin l$, Q' is the fourth corner of the parallelogram on P, P', Q , and we set $\tau_{PP'}(Q) = Q'$.

Step 2. If $\tau_{PP'}(Q) = Q'$, then for any $R \notin PP'$, and $R \notin QQ'$, we have

$$\tau_{PP'}(R) = \tau_{QQ'}(R) .$$

Indeed, define

$$R' = \tau_{PP'}(R) .$$

Then by A5a , $QR \parallel Q'R'$, so we have also

$$R' = \tau_{QQ'}(R) .$$

Step 3. Starting with P, P', Q , taking $Q' = \tau_{PP'}(Q)$, we can now define τ to be $\tau_{PP'}$ or $\tau_{QQ'}$, whichever one happens to be defined at a given point, since we saw they agree where they are both defined.

Step 4. Note that if R is any point, and $\tau(R) = R'$, then $\tau = \tau_{RR'}$ whenever they are both defined. This follows as above.

Step 5. Clearly τ is 1-1 and onto. If X, Y, Z are collinear point, let X', Y', Z' be their images. Then

$$\tau(Y) = \tau_{XX'}(Y)$$

and

$$\tau(Z) = \tau_{XX'}(Z).$$

So it follows immediately from the definition of $\tau_{XX'}$ that X', Y', Z' are collinear. Hence τ is an automorphism of A. One sees immediately from the construction that it is a dilatation with no fixed points, hence is a translation, and it takes P to P'.

Finally, the uniqueness of τ follows from the fact that a translation with a fixed point is the identity.

(ii) $\Rightarrow$ (i). We assume the existence of translations, and must deduce A5a. Suppose given l, m, n, A, A', B, B', C, C', as in the statement of A5a, and let τ be a translation taking A into A'. Then by our hypotheses, $\tau(B) = B'$ and $\tau(C) = C'$. Hence $BC \parallel B'C'$ since τ is a dilatation.

Proposition 7.7. (Assuming A5a) Tran(A) is an abelian group.

Proof: Let τ, τ' be translations. We must show $\tau\tau' = \tau'\tau$.

Case 1. τ and τ' translate in different directions. Let P be a point. Let $\tau(P) = P'$, $\tau'(P) = Q$. Then

$$\tau(Q) = \tau\tau'(P)$$

and

$$\tau'(P') = \tau'\tau(P)$$

are both found as the fourth vertex of the parallelogram on P, P', Q, hence are equal, so $\tau\tau' = \tau'\tau$. (Note so far we have not used A5a).

Case 2. τ and τ' are in the same direction. Let τ^* be a translation in a different direction (here we use Theorem 7.6 and axiom A 3 to ensure that there is another direction, and a translation in that direction.) Then

$$\tau\tau' = \tau\tau'\tau^*\tau^{*-1} = (\tau'\tau^*)\tau\tau^{*-1}$$

since τ and $\tau'\tau^*$ are in different directions. This equals

$$\tau'\tau\tau^*\tau^{*-1} = \tau'\tau \ .$$

Since τ and τ^* are in different directions.

q. e. d.

Definition. Let G be a group, and let H, K be subgroups. We say G is the semi-direct product of H and K if

1) H is a normal subgroup of G

2) $H \cap K = \{1\}$

3) H and K together generate G.

This implies that every element $g \in G$ can be written uniquely as a product $g = hk$, $h \in H$, $k \in K$.

Definition. Let O be a point in A, and define $\mathrm{Dil}_O(A)$ to be the subgroup of Dil(A) consisting of those dilatations φ such that $\varphi(O) = O$.

Proposition 7.8 Dil(A) is the semi-direct product of Tran(A) and $\mathrm{Dil}_O(A)$.

Proof : 1) we have seen that Tran(A) is a normal subgroup of Dil(A) .

2) If $\varphi \in$ Tran(A) $\cap$ $Dil_o(A)$, then φ has a fixed point, but being a translation it must be the identity.

3) Let $\varphi \in$ Dil(A) . Let $\varphi(O) = Q$.
Let τ be a translation such
that $\tau(O) = Q$. Then $\tau^{-1} \varphi \in Dil_o(A)$, so $\varphi = \tau . \tau^{-1} \varphi$ shows that Tran(A) and $Dil_o(A)$ generate Dil(A) . Note here we have used the existence of translations.

A 5 a (Big Desargues' Axiom) Let O, A, B, C, A', B', C' be distinct points in the affine plane A, and assume that

O, A ,A' are collinear

O, B, B' are collinear

O, C, C' are collinear

AB $\|$ A'B'

AC $\|$ A'C'

Then BC $\|$ B'C' .

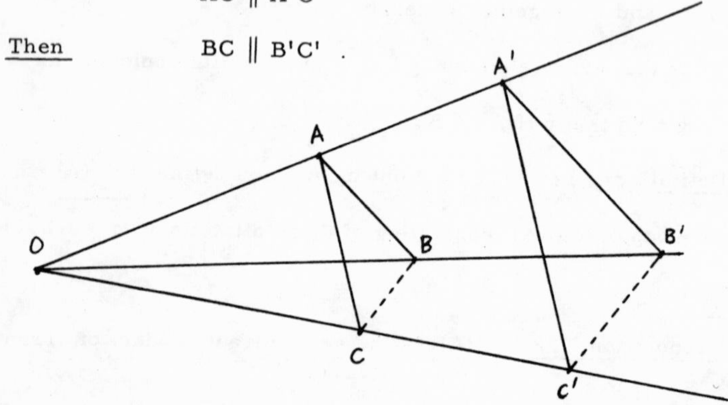

Note that this statement follows from P 5 , if A is embedded in a projective plane π .

<u>Theorem 7.9.</u> The following two statements are equivalent, in the affine plane A.

(i) The axiom A5b holds in A

(ii) Given any three points O, P, P' , with $P \neq O$, $P' \neq O$, and O, P, P' are collinear, there exists a unique dilatation σ of A, such that $\sigma(O) = O$ and $\sigma(P) = P'$.

<u>Proof :</u> The proof is entirely analogous to the proof of theorem 7.6 , so the details will be left to the reader. Here is an outline :

(i) ⇒ (ii) . Given O, P, P' , as above, define a transformation $\varphi_{O,P,P'}$, for points Q not on the line l containing O, P, P'

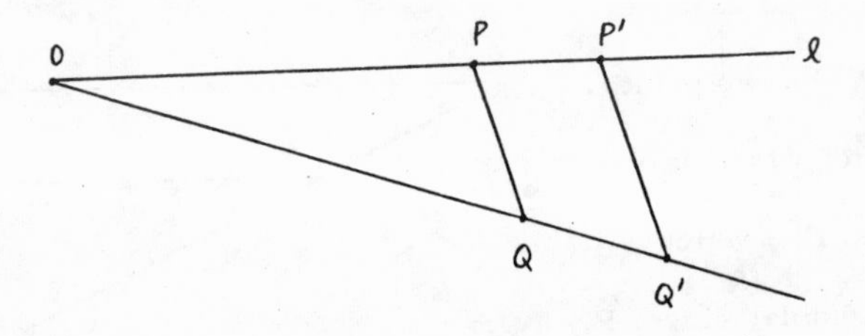

as follows . $\varphi_{O,P,P'}(Q) = Q'$, where Q' is the intersection of the line OQ with the line through P' , parallel to PQ.

Now if $\varphi_{O,P,P'}(Q) = Q'$, one proves using A5b that $\varphi_{O,P,P'}$ agrees with $\varphi_{O,Q,Q'}$ (defined similarly) whenever both are defined. Hence one can define σ to be either one, and $\sigma(O) = O$. Then σ

is defined everywhere. Next show that if $\sigma(R) = R'$, $R \neq O$, then $\sigma = \varphi_{O,R,R'}$ whenever the latter is defined. Now clearly σ is 1 - 1 and onto. But using the previous results, one can show easily that it takes lines into lines, so is an automorphism, and that $PQ \parallel \sigma(P)\sigma(Q)$ for any P, Q, so σ is a dilatation. The uniqueness follows from Corollary 7.3.

(ii) $\Rightarrow$ (i). Let O, A, B, C, A', B', C' be given satisfying the hypotheses of A5b. Let σ be a dilatation which leaves O fixed and sends A into A'. Then by the hypotheses, $\sigma(B) = B'$, and $\sigma(C) = C'$. So from the fact that σ is a dilatation, $BC \parallel B'C'$

Remark: Using the theorems 7.6 and 7.9, we can show that A5b $\Rightarrow$ A5a, although this is not obvious from the geometrical statements.

Indeed, let us assume A5b. Let P, P' be two points. We will construct a translation sending P into P', which will show that A5a holds, since P, P' are arbitrary.

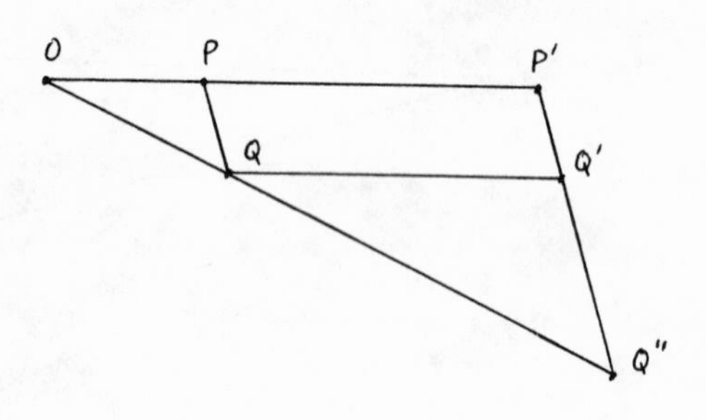

Let Q be a point not on PP', and let Q' be the fourth vertex of the parallelogram on P, P', Q. Let O be a point on PP', $\neq P$, and $\neq P'$. Let σ_1 be a dilatation which leaves O fixed, and sends P to P' (which exists by theorem 7.9). Let $\sigma_1(Q) = Q''$. Then P', Q', Q'' are collinear, so there exists a dilatation σ_2 leaving P' fixed, and sending Q'' to Q'.

Now consider $\tau = \sigma_2\sigma_1$. Being a product of dilatations, it is itself a dilatation. One sees easily that $\tau(P) = P'$ and $\tau(Q) = Q'$. Now any fixed point of τ must lie on PP' and on QQ' (because if X is a fixed point $XP \parallel XP' \Rightarrow X, P, P'$ collinear , similar for Q). But $PP' \parallel QQ'$, so τ has no fixed points (we are implicitly assuming $P \neq P'$: but if $P = P'$ we could have taken the identiy, which is a translation sending P to P). Hence τ is a translation sending P into P' , so by theorem 7.6 , A5a holds.

Now we come to the construction of coordinates in the affine plane A. In fact, we will find it convenient to construct a few more things, while we are at it. So our program is to construct the following objects :

1) We will define a division ring F.

2) We will assign coordinates to the points of A, so that A is in 1-1 correspondence with the set of ordered pairs of elements of F.

3) We will find the equation of an arbitrary translation of A, in terms of the coordinates

4) We will find the equation of an arbitrary dilatation.

5) Finally, we will show that the lines in A are given by linear equations, and this will prove that A is isomorphic to the affine plane $\mathbb{A}^2_F$.

In the course of these constructions, there will be about a thousand details to verify, so we will not attempt to do them all, but will give indications, and leave the trivial verifications to the reader.

1) Definition of F.

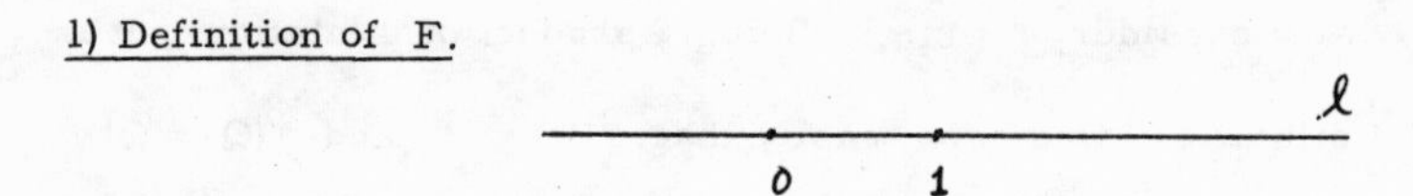

Fix a line l in A, and fix two points on l, call them o, 1 . Now let F be the set of points of l.

If $a \in F$, (i.e. if a is a point of l) , let τ_a be the unique translation which takes o into a (here we use A5a) . If $a \in F$ and $a \neq o$, let σ_a be the unique dilatation of A which leaves o fixed , and sends 1 into a.

Now we define addition and multiplication in F as follows.

If $a, b \in F$, define

$$a + b = \tau_a \tau_b (o) = \tau_a (b).$$

Since the translations form an abelian group, we see immediately that addition is associative and commutative :

$$(a+b) + c = a + (b + c)$$

$$a + b = b + a ,$$

and that o is the identity element, and that $\tau_a^{-1}(o) = -a$ is the additive inverse. Thus F is an abelian group under addition. (Notice how much simpler these verifications are than if we had followed the plan suggested on pp. 96 - 97.)

Note also from our definition of addition that we have

$$\tau_{a+b} = \tau_a \tau_b \qquad \text{for all } a, b \in F \quad .$$

Now we define multiplication as follows : o times anything is o.

If $a, b \in F$, $b \neq o$, we define

$$ab = \sigma_b(a) = \sigma_b \sigma_a(1) .$$

Now since the dilatations form a group, we see immediately that

$$(ab)c = a(bc) ,$$

$$a \cdot 1 = 1 \cdot a = a \qquad \text{for all } a$$

$$\sigma_a^{-1}(1) = a^{-1} \qquad \text{is a multiplicative inverse.}$$

Therefore the non-zero elements of F form a group under multiplication. Furthermore, we have the formulae (for $b \neq o$)

$$\tau_{ab} = \sigma_b \tau_a \sigma_b^{-1}$$

$$\sigma_{ab} = \sigma_b \sigma_a$$

It remains to establish the distributive laws in F. For some reason, one of them is much harder than the other, perhaps because our definition of multiplication is asymmetric. First consider $(a + b).c$. If $c = o$, $(a + b)c = o = ac + bc$ ok. If $c \neq o$, we use the formulae above, and find

$$\tau_{(a+b)c} = \sigma_c \tau_{a+b} \sigma_c^{-1} = \sigma_c \tau_a \tau_b \sigma_c^{-1} = \sigma_c \tau_a \sigma_c^{-1} \sigma_c \tau_b \sigma_c^{-1} =$$

$$\tau_{ac} \tau_{bc} = \tau_{ac+bc} .$$

Now applying both ends of this equality to the point o, we have

$$(a + b)c = ac + bc.$$

Before proving the other distributivity law, we must establish a lemma. For any line m in A, let $\underline{Tran_m(A)}$ be the group of translations in the direction of m, i.e. those translations $\tau \in Tran(A)$

such that either $\tau = \text{id.}$ or $PP' \parallel m$ for all P (Where $\tau(P) = P'$) .

Lemma 7.10 Let m, n be lines in A (which may be the same) .

Let $\tau' \in \text{Tran}_m(A)$ and $\tau'' \in \text{Tran}_n(A)$ be fixed translations, different from the identiy, and let o be a fixed point of A. We define a mapping

$$\varphi : \text{Tran}_m(A) \longrightarrow \text{Tran}_n(A)$$

as follows : For each $\tau \in \text{Tran}_m(A)$, $\tau \neq \text{id.}$, there exists a unique dilatation $\sigma \in \text{Dil}_o(A)$, leaving o fixed, and such that

$$\tau = \sigma\tau'\sigma^{-1}$$

(Indeed, take σ such that $\sigma(\tau'(o)) = \tau(o)$.) Define

$$\varphi(\tau) = \sigma\tau''\sigma^{-1}$$

(with that σ) .

Then , φ is a homomorphism of groups, i. e. for all $\tau_1, \tau_2 \in \text{Tran}_m(A)$, $\varphi(\tau_1\tau_2) = \varphi(\tau_1)\varphi(\tau_2)$.

Proof : Case 1.

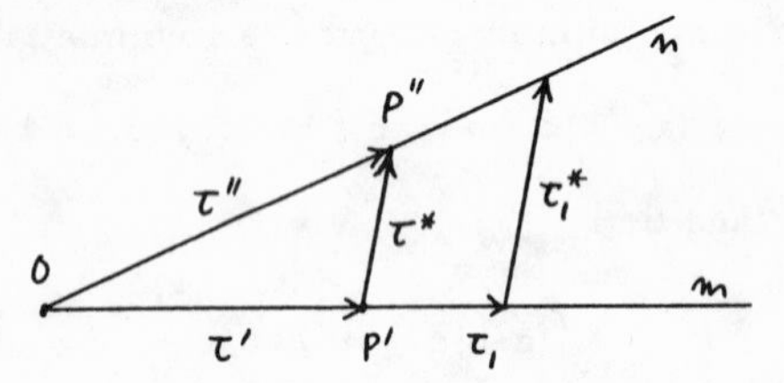

First we treat the case where $m \nparallel n$. Replacing m, n by lines parallel to them, if necessary, we may assume that m and n pass through O . Let $\tau'(o) = P'$, $\tau''(o) = P''$.

Let τ^* be the unique translation which takes P' into P'' . Then

$$\tau'' = \tau'\tau^*.$$

If $\tau_1, \tau_2 \in \text{Tran}_m(A)$, let σ_1, σ_2 be the corresponding dilatations. Then

$$\varphi(\tau_1) = \sigma_1\tau''\sigma_1^{-1} = \sigma_1\tau'\tau^*\sigma_1^{-1} = \sigma_1\tau'\sigma_1^{-1}\sigma_1\tau^*\sigma_1^{-1}$$

$$= \tau_1 \cdot \sigma_1\tau^*\sigma_1^{-1} = \tau_1\tau_1^* ,$$

where we define

$$\tau_1^* = \sigma_1 \tau^* \sigma_1^{-1} .$$

Similarly,

$$\varphi(\tau_2) = \tau_2 \tau_2^* ,$$

where

$$\tau_2^* = \sigma_2 \tau^* \sigma_2^{-1} ,$$

and

$$\varphi(\tau_1\tau_2) = \tau_1\tau_2 \cdot \tau_3^* ,$$

where σ_3 corresponds to τ_1, τ_2 , and

$$\tau_3^* = \sigma_3 \tau^* \sigma_3^{-1} .$$

So we have

$$\varphi(\tau_1\tau_2) = \tau_1\tau_2 \cdot \tau_3^*$$

$$\varphi(\tau_1)\varphi(\tau_2) = \tau_1\tau_2 \cdot \tau_1^*\tau_2^* .$$

Now $\varphi(\tau_1\tau_2)$ and $\varphi(\tau_1)\varphi(\tau_2)$ are both translations in the m direction. τ_3^* and $\tau_1^*\tau_2^*$ are both translations in the τ^* direction. But this can only happen if

$$\tau_3^* = \tau_1^*\tau_2^*$$

and

$$\varphi(\tau_1\tau_2) = \varphi(\tau_1)\varphi(\tau_2),$$

which is what we anted to prove. (To make this argument more explicit consider the points Q and R, which are the images of O under the two translations above. Then we have O, Q, R collinear, and also $\tau_1\tau_2(o)$, Q, R collinear which implies Q = R.)

<u>Case 2.</u> If $m \parallel n$, τ' , $\tau'' \in \mathrm{Tran}_m(A)$. Take another line o, not parallel to m, and take $\tau''' \in \mathrm{Tran}_o(A)$. Define

$$\psi_1 : \mathrm{Tran}_m(A) \longrightarrow \mathrm{Tran}_o(A)$$

using τ' and τ''' , and define

$$\psi_2 : \mathrm{Tran}_o(A) \longrightarrow \mathrm{Tran}_m(A)$$

using τ''' and τ'' .

Then $\varphi = \psi_2\psi_1$, and ψ_1, ψ_2 are homomorphism by Case 1, so φ is a homomorphism.

(Note the analogy of this proof with the proof of Proposition 7.7)

q. e. d.

Now we can prove the other distributivity law , as follows. Consider $\lambda(a + b)$. In the lemma, take $m = n = 1$. $o = o, \tau' = \tau_1$, $\tau'' = \tau_\lambda$. Then φ is the map of $\mathrm{Tran}_1(A) \longrightarrow \mathrm{Tran}_1(A)$ which sends τ_a into $\tau_{\lambda a}$, for any a. Indeed, $\tau_a = \sigma_a\tau_1\sigma_a^{-1}$, so $\sigma = \sigma_a$ and $\sigma_a\tau_\lambda\sigma_a^{-1} = \tau_{\lambda a}$. Now the lemma tells us that φ is a homomorphism, i. e. for any $a, b \in F$,

$$\varphi(\tau_a\tau_b) = \varphi(\tau_a)\varphi(\tau_b)$$

or

$$\varphi(\tau_{a+b}) = \varphi(\tau_a)\varphi(\tau_b) .$$

Hence

$$\tau_{\lambda(a+b)} = \tau_{\lambda a}\tau_{\lambda b} = \tau_{\lambda a+\lambda b} .$$

Applying both sides to o, we have

$$\lambda(a+b) = \lambda a + \lambda b .$$

Thus we have proved

<u>Theorem 7.11</u> Let A be an affine plane satisfying A5a and A5b. Let l be a line of A, let o, 1 be two points of l, let F be the set of points of l, and define + and . in F as above. Then F is a division ring.

Now we can introduce coordinates in A. We have already fixed a line l in A, and two points o, 1 on l, and on the basis of these choices we defined our divising ring F. No we choose another line, m, passing through o, and fix a point 1' on m.

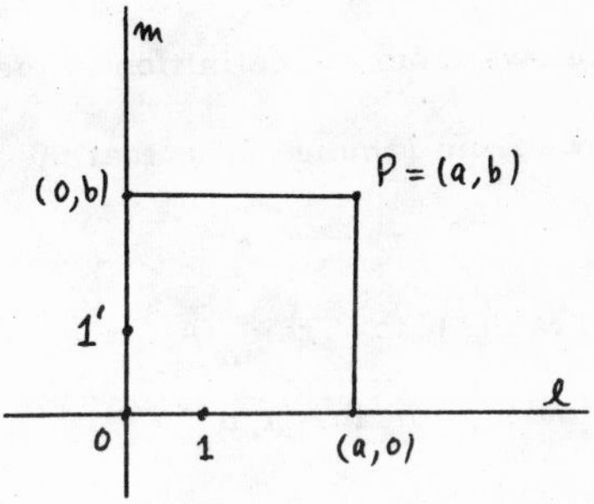

For each point $P \in l$, if P corresponds to the element $a \in F$, we give P the coordinates (a, o). Thus o and 1 have coordinates (o, o) and (1, o), respectively.

If $P \in m$, $P \neq o$, then there is a unique dilatation σ leaving o fixed and sending 1' into P. σ must be of the form σ_a for some $a \in F$. So we give P the coordinates (o, a).

Finally if P is a point not on l or m, we draw lines through P, parallel to l and m, to intersect m in (o, b) and l in (a, o). Then we give P the coordinates (a, b).

One sees easily that in this way A is put into 1-1 correspondence with the set of ordered pairs of elements of F. We have yet to see that lines are given by linear equations - this will come after we find the equations of translations and dilatations.

Now we will investigate the equations of translations and dilatations. First some notation. For any $a \in F$, denote by τ'_a the translation which takes o into (o, a). Thus τ'_1 is the translation which takes o into 1', and for any $a \in F$, $a \neq o$,

$$\tau'_a = \sigma_a \tau'_1 \sigma_a^{-1} .$$

This follows from the definition of the point (o, a). Furthermore, it follows from lemma 7.10 that the mapping

$$\tau_a \longrightarrow \tau'_a$$

from $\mathrm{Tran}_l(A)$ to $\mathrm{Tran}_m(A)$ is a homomorphism, and hence we have the formulae, for any $a, b \in F$,

$$\tau'_{a+b} = \tau'_a \tau'_b$$

$$\tau'_{ab} = \sigma_b \tau'_a \sigma_b^{-1}$$

<u>Proposition 7.12</u>. Let τ be a translation of A, and suppose that $\tau(o) = (a, b)$. Then τ takes an arbitrary point $Q = (x, y)$ into $Q' = (x', y')$ where

$$\begin{cases} x' = x + a \\ y' = y + b . \end{cases}$$

<u>Proof</u>: Indeed, let τ_{oQ} be the translation taking o into Q. Then $\tau_{oQ} = \tau_x \tau'_y$. Also $\tau = \tau_a \tau'_b$. So

$$\tau(Q) = \pi_{oQ}(o) =$$

$$= \tau_a \tau'_b \tau_x \tau'_y(o) = \tau_a \tau_x \tau'_b \tau'_y(o)$$

$$= \tau_{a+x} \tau'_{b+y}(o) = (x + a,\ y + b)\ .$$

Proposition 7.13 Let σ be any dilatation of A leaving o fixed. Then $\sigma = \sigma_a$ for some $a \in F$, and σ takes the point $Q = (x, y)$ into $Q' = (x', y')$, where

$$\begin{cases} x' = xa \\ y' = ya\ . \end{cases}$$

Proof : Again write $\tau_{oQ} = \tau_x \tau'_y$. Then

$$\sigma(Q) = \sigma_a \tau_x \tau'_y(o) = \sigma_a \tau_x \tau'_y \sigma_a^{-1}(o)$$

$$= \sigma_a \tau_a \sigma_a^{-1} \cdot \sigma_a \tau'_y \sigma_a^{-1}(o)$$

$$= \tau_{xa} \cdot \tau'_{ya}(o) = (xa, ya)\ .$$

Theorem 7.14. Let A be an affine plane satisfying A5a and A5b. Fix two lines l, m in A, and fix points $1 \in l$, $1' \in m$, different from $o = l.m$. Then assigning coordinates as above, the lines in A are given by linear equations of the form

$$y = mx + b \qquad\qquad m, b \in F$$

or

$$x = a \qquad\qquad a \in F.$$

Thus A is isomorphic to the affine plane $\mathbb{A}^2_F$.

Proof : By construction of the coordinates, a line parallel to l will have an equation of the form $y = b$, and a line parallel to m will have an equation of the form $x = a$.

Now let r be any line through o, different from l and m.

Then r must intersect the line $x = 1$, say in the point $Q = (1, m)$ $(m \in F)$.

Now if R is any other point on r, different from o, there is a unique dilatation σ_λ leaving o fixed and sending Q into R. Hence R will have coordinates

$$x = 1 \cdot \lambda$$

$$y = m \cdot \lambda .$$

Eliminating λ, we find the equation of r is

$$y = mx.$$

Finally, let s be a line not passing through o, and not parallel to l or m. Let r be the line parallel to s, passing through o. Let s intersect m in (o, b).

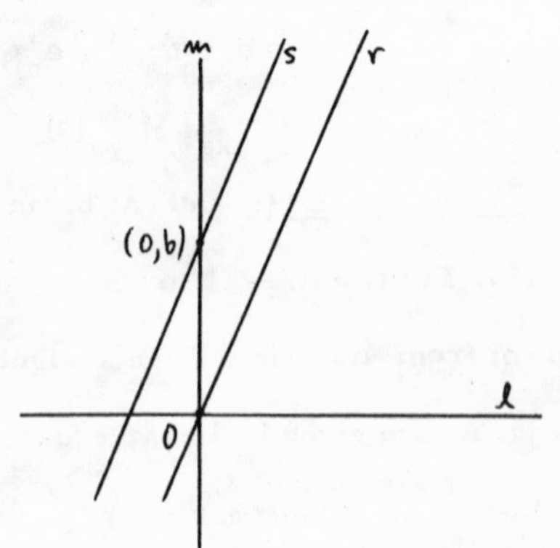

Then it is clear that the points of s are obtained by applying the translation τ'_b to the points of r. So if $(\lambda, m\lambda)$ is a point of r (for $x = \lambda$), the corresponding point of s will be

$$x = \lambda + o$$

$$y = m\lambda + b$$

So the equation of r is

$$y = mx + b$$

q. e. d.

Remark : If σ is an arbitrary dilatation of A, then σ can be written as $\tau\sigma'$, where τ is a translation , and σ' is a dilatation leaving o fixed (cf. Proposition 7.8) . So if τ has equations

$$\begin{cases} x' = x + c \\ y' = y + d \end{cases}$$

and σ' has equations

$$\begin{cases} x' = xa \\ y' = ya , \end{cases}$$

we find that σ has equations

$$\begin{cases} x' = xa + c \\ y' = ya + d . \end{cases}$$

Theorem 7.15 . Let π be a projective plane satisfying P 1- P 5 . Then there is a division ring F such that π is isomorphic to $\mathbb{P}^2_F$, the projective plane over F.

Proof : Let l_o be any line in π , and consider tha affine plane $A = \pi - l_o$. Then A satisfies A5a and A5b , hance $A \cong \mathbb{A}^2_F$, by the previous theorem. But π is the projective plane associated to the affine plane A, and $\mathbb{P}^2_F$ is the projective plane associated to the affine plane $\mathbb{A}^2_F$, so this isomorphism extends to show $\pi \cong \mathbb{P}^2_F$.

Remark : This is a good point to clear up a question left hanging from chapter 1, about the correspondence between affine planes and projective planes. We saw that an affine plane A could be completed to a projective plane S(A) by adding ideal points and an ideal line. Conversely, if π is a projective plane, and l_o a line in π then $\pi - l_o$ is an affine plane.

What happens if we perform first one process and then the other ? Do we get back where we started ? There are two cases to consider.

1) If π is a projective plane, l_o a line in π, $\pi - l_o$ the corresponding affine plane, then one can see easily that $S(\pi - l_o)$ is isomorphic to π in a natural way.

2) Let A be an affine plane, and let $S(A) = A \cup l_\infty$ be the corresponding projective plane. Then clearly $S(A) - l_\infty \cong A$. But suppose l_1 is a line in $S(A)$, different from l_∞ ? Then in general one cannot expect $S(A) - l_1$ to be isomorphic to A.

For example, let Π be the free projective plane on the configuration π_o = a projective plane on seven points, plus one more point. Let $A = \Pi - l_\infty$, where l_∞ is one of the lines of π_o. Then $S(A) = \Pi$. Let l_1 be a line of Π containing no point of π_o. Then $\Pi - l_1$ is not isomorphic to A, because $\Pi - l_1$ contains a confined configuration, but A contains no confined configuration.

However, if we assume that A satisfies A5a and A5b, then $S(A) - l_1 \cong A$. Indeed, $S(A) \cong \mathbb{P}^2_F$, for some division ring F, and we can always find an automorphism $\varphi \in \mathrm{Aut}\, \mathbb{P}^2_F$, taking l_1 to l_∞. (see Proposition 6.6) Then φ gives an isomorphism of $S(A) - l_1$ and A.

CHAPTER 8. PROJECTIVE COLLINATIONS

Let us look back for a moment at what we have accomplished so far. We have been approaching the subject of projective geometry from two different directions, the synthetic and the analytic.

The synthetic approach starts from the axioms P 1 - P 4, and eventually P 5, P 6, P 7, and builds everything in logical steps from there. Thus we have the notion of harmonic points, of perspectivities and projectivities from one line to another, and the Fundamental Theorem, which says that there is a unique projectivity from a line l into itself which sends three given points $A, B, C,$ into three other given points A', B', C'.

The analytic approach starts from an algebraic object, such as a division ring or field F, or the real numbers $\mathbb{R}$. Then we define $\mathbb{P}^2_F$ as triples of elements of the field with a certain equivalence relation, and lines as linear equations. We can define certain automorphisms of $\mathbb{P}^2_F$ using matrices, others using automorphisms of F, and we have a Fundamental theorem telling us that these two types of automorphisms generate the entire group of automorphisms of $\mathbb{P}^2_F$.

In the last two chapters, we have tied these two approaches together, by showing that a (synthetic) projective plane is of the form $\mathbb{P}^2_F$ for some division ring F, if an only if Desargues' axiom, P 5, holds. Furthermore, we showed that the axioms P 6, and P 7, which are synthetic statements, are equivalent to algebraic statements about the division ring F.

In this chapter we will continue exploring the relationship between the synthetic and the analytic approaches, in two important situations. One is to give an analytic interpretation of the group $PJ(l)$ of projectivities of a line into itself, which so far we have studied only from the synthetic point of view. The other is to give a synthetic interpretation of the group $PGL(2)$ of automorphisms of $\mathbb{P}^2_F$ defined by matrices, which so far we have studied only from the analytic point of view.

Projectivities on a line.

Let F be a field (we will stick to the commutative case for simplicity), and let $\pi = \mathbb{P}^2_F$ be the projective plane over F. Then π satisfies P 5 and P 6. Let l be the line $x_3 = o$, so that l has homogeneous coordinates x_1 and x_2. We have already studied the group $PJ(l)$ of projectivities of l into itself (see Chapter 5). Now we will define another group of transformations of l into itself, $PGL(l)$, and will prove it is equal to $PJ(l)$.

Let $A = \begin{pmatrix} a & b \\ c & d \end{pmatrix}$ be a 2×2 matrix with coefficients in F, and with $\det A \equiv ad - bc \neq o$. Then we define a transformation of l into itself by the equations

$$x_1' = ax_1 + bx_2$$
$$x_2' = cx_1 + dx_2 .$$

Call this transformation T_A. As in chapter 3, one can show easily that T_A is a one - to - one tranformation of 1 onto itself, whose inverse is $T_{A^{-1}}$. If A, B are two such matrices, then $T_A T_B = T_{AB}$, so the set of all such transformations forms a group. Two matrices A and A' define the same transformation (i. e. $T_A = T_{A'}$) if and only if there is an element $\lambda \in F$, $\lambda \neq o$, such that $A' = \lambda A$.

Definition. The group of transformations of 1 into itself of the form T_A defined above, where $A = \begin{pmatrix} a & b \\ c & d \end{pmatrix}$ is a matrix of elements of F, with $ad - bc \neq o$, is called PGL(1;F), or PGL(1) for short.

In dealing with the group PGL(1), we will find it more convenient to introduce a non-homogeneous coordinate $x = x_1/x_2$ on 1. Thus x may take on all values of F, plus the value ∞ (where $a/o = \infty$. for any $a \in F$, $a \neq o$). Then the points of 1 are in one-to-one correspondence with the elements of the set $F \cup \{\infty\}$. Furthermore, the group PGL(1) is then the group of fractional linear transformations of 1, namely those given by equations of the form

$$x' = \frac{ax + b}{cx + d} \qquad ad - bc \neq o, \ a, b, c, d \in F.$$

When $x = \infty$, this expression is defined to be a/c, if $c \neq o$, and ∞ if $c = o$ (note that $a = c = o$ is impossible becuase of the condition $ad - bc \neq o$).

Proposition 8.1. Let A, B, C and A', B', C' be two triples of distinct points on 1. Then there is a unique element of PGL(1) which sends A, B, C, into A', B', C', respectively.

Proof : The proof could be done as in Chapter 3 for PGL(2), but it is simple enough to be worth repreating in this new context.

For the existence of such a transformation, it is sufficient to consider the case where $A, B, C = o, 1, \infty$, respectively, and where A', B', C', are three points with coordinates α, β, γ respectively. Then we must find a, b, c, d so that the transformation

$$x' = \frac{ax + b}{cx + d}$$

takes $o, 1, \infty$ to α, β, γ. So we must solve

$$\alpha = \frac{b}{d} \ , \quad \beta = \frac{a + b}{c + d} \ , \quad \gamma = \frac{a}{c} \quad .$$

Suppose that α, β, γ are all different form ∞. (We leave the special case when one of them is ∞ to the reader !) . Then set $d = 1$, and solve the other equations, finding

$$b = \alpha \ , \quad c = \frac{\alpha - \beta}{\beta - \gamma} , \qquad a = \frac{\alpha - \beta}{\beta - \gamma} \ . \ \gamma \quad .$$

Then

$$ad - bc = \frac{\alpha - \beta}{\beta - \gamma} (\gamma - \alpha) \neq o$$

since α, β, γ are all distinct . Thus we have a transformation of the right kind, which does what we want.

To show uniqueness, it is sufficient to show that if the transformation

$$x' = \frac{ax + b}{cx + d}$$

leaves $o, 1, \infty$ fixed, then it is the identity. Indeed, in that case we have

$$o = \frac{b}{d} \ , \quad 1 = \frac{a + b}{c + d} \ , \quad \infty = \frac{a}{c} \ ,$$

which implies $b = o$, $c = o$, $a = d$, so $x' = x$.

Proposition 8.2. The group PGL(1) of fractional linear transformations is generated by transformations of the following three kinds :

(i) $x' = x + a$ $\quad a \in F$

(ii) $x' = ax$ $\quad a \in F, \; a \neq o$

(iii) $x' = \frac{1}{x}$.

(each of which is , of course, a fractional linear transformation).

Proof : First of all, it is clear that by using a type (ii) transformation, followed by a type (i) tranformation, we can get an arbitrary transformation of the form (*)

$$x' = ax + b \qquad a, b \in F, \; a \neq o.$$

Now let

$$x' = \frac{ax + b}{cx + d} \qquad ad - bc \neq o$$

be an arbitrary fractional linear equations, If $c = o$, then $x' = \frac{a}{d} x + \frac{b}{d}$ = and $\frac{a}{d} \neq o$, so it is the above form (*) . So we may suppose $c \neq o$. Then let $x_1 = cx + d$, so that $x = \frac{1}{c}(x_1 - d)$ and

$$x' = \frac{a\frac{1}{c}(x_1 - d) + b}{x_1} = \frac{b - \frac{ad}{c}}{x_1} + \frac{a}{c} .$$

Now $b - \frac{ad}{c} \neq o$ by hypothesis, hence x' can be obtained from x_1 by an application of (iii) followed by one of the above type (*) .

Thus all together, x' is obtained form x by one application of (iii) , and two applications each of transformations of the types (ii) and (i) .

Proposition 8.3. Each one of the three special types of transformations (i) , (ii) , and (iii) of the previous proposition is a projectivity of 1 into itself .

Proof : We must exhibit each of these transformations as a product of perspectivities, to show that it is a projectivity.

(i) $x' = x + a$.

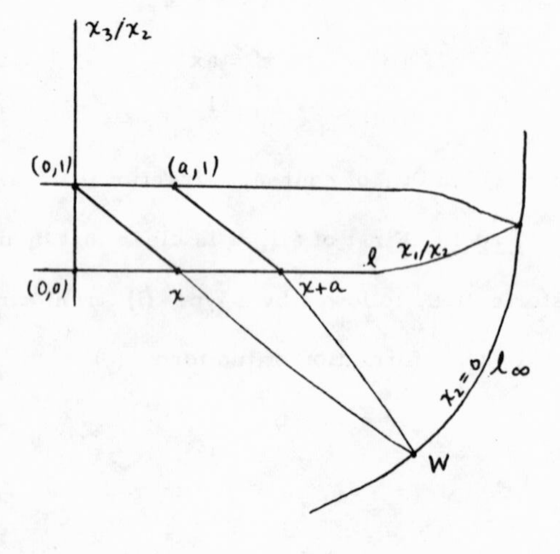

Take $x_2 = o$ to be the line at infinity, and take affine coordinates $x = x_1/x_2$, $y = x_3/x_2$ in the affine plane. Then 1 is the x-axis, and we can construct $x + a$ geometrically as follows :

1) Project (x, o) from the point $(o, 1)$ onto the line l_∞, getting W.

2) Project W back onto 1 from the point $(a, 1)$. This gives $x + a$.

Thus the transformation $x' = x + a$ is a product of two perspectivities, and so is a projectivity.

(ii) $x' = ax$, $a \neq o$.

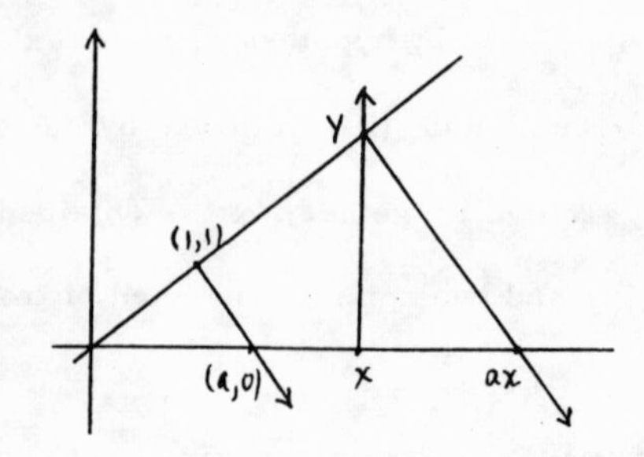

This transformation, too, is a product of two perspectivities.

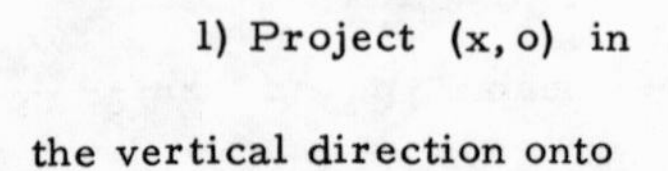

1) Project (x, o) in the vertical direction onto

the line $x = y$, getting the point Y.

2) Project Y back onto l, in the direction of the line joining (1, 1) and (a, o) to obtain the point (ax, o).

(iii) $x' = \frac{1}{x}$

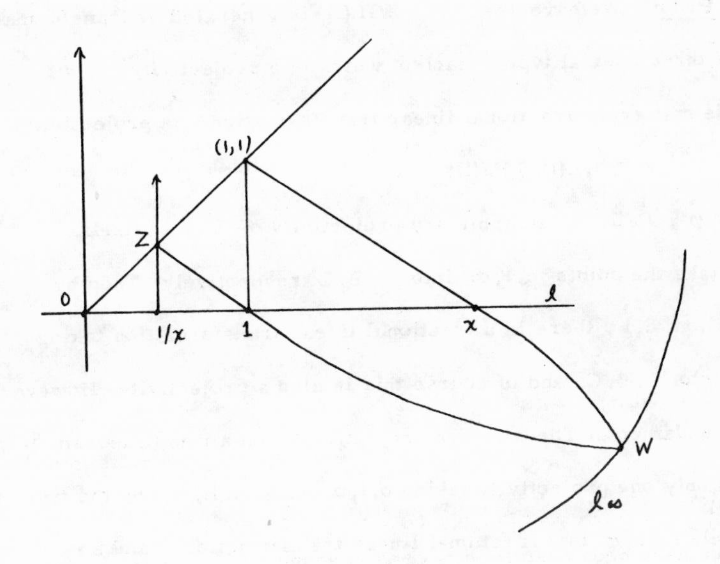

This transformations is a product of three perspectivities.

1) Project (x, o) from the point (1, 1) onto the line at infinity, l_∞, getting W.

2) Project W from the point (1, o) onto the line $x = y$, getting Z.

3) Project Z in the vertical direction back onto l, getting the point $(\frac{1}{x}, o)$.

q. e. d.

<u>Theorem 8.4.</u> Let F be a field, let $\pi = \mathbb{P}^2_F$, let l be the line $x_3 = o$. Then the group PJ(l) of projectivities of l into itself is equal to the group PGL(l) of fractional linear transformations on l.

<u>Proof :</u> We have seen that PGL(l) is generated by transformations of three special types, each of which is a projectivity. So we conclude that every fractional linear transformation is a projectivity, i. e.

$$PGL(l) \subseteq PJ(l).$$

Now let $\varphi \in PJ(l)$ be an arbitrary projectivity of l into itself. Let φ take the points $o, 1, \infty$ into A, B, C respectively. Then by Proposition 8.1 , there is a fractional linear transformation taking $o, 1, \infty$ into A, B, C, and of course this is also a projectivity. However, by the Fundamental Theorem for projectivities on a line (Theorem 5.6) there is only one projectivity taking $o, 1, \infty$ into A, B, C. So the two are equal, i. e. φ is a fractional linear transformation , and so

$$PGL(l) = PJ(l) .$$

q. e. d.

<u>Remarks:</u> 1. Notice that we have had to use the full strength of our synthetic theory (in the form of the Fundamental Theorem for projectivities on a line, which was a hard theorem) to prove this result. And that is not surprising , because what we have proved is really a rather remarkable fact. It says that our two entirely different approaches have actually converged, and that we have arrived in each case at the

same group of transformations of the line into itself.

2. One may wonder what is special about the line $x_3 = o$ which occurs in the statement of the theorem. Nothing is special about it. More precisely, if l' is any other line, then the groups $PJ(l)$ and $PJ(l')$ are isomorphic, as abstract groups. To get such an isomorphism, let P be any point not on l or l', and let $\psi : l \longrightarrow l'$ be the perspectivity $l \overset{P}{\underset{\wedge}{=}} l'$. Then for each $\alpha \in PJ(l)$, we have $\psi\alpha\psi^{-1} \in PJ(l')$, and the mapping

$$\alpha \longmapsto \psi\alpha\psi^{-1}$$

is an isomorphism of $PJ(l)$ onto $PJ(l')$. (Details left to the reader !). One will note, however, that this isomorphism depends on the choice of P. In fact, there is no one way to make $PJ(l)$ and $PJ(l')$ isomorphic, which is better than all other ways. So we say $PJ(l)$ and $PJ(l')$ are <u>non-canonically</u> isomorphic

To recapitulate, we have been examining a certain group of transformations of the line l into itself, namely $PJ(l) = PGL(l)$ and have found that we can describe it in two different ways. One is by considering l as a line in $\mathbb{P}^2_F$, and using incidence properties of the projective plane. The other is by using the algebraic sturcture on l given by its coordinatization. Now we will give a third way of characterizing these transformations, namely as the group of all permutations of l which preserve cross-ratio (this notion will be explained presently) . Finally , in case F is the field $\mathbb{C}$ of complex numbers, we will give a fourth interpretation of this group, as the group of all conformal, orientation-preserving maps of the Riemann sphere onto itself.

Definition. Let F be a field, and let a, b, c, d be four distinct points on the line l as above, i.e. $a, b, c, d \in F \cup \{\infty\}$. Then we define the cross-ratio of the four points, by

$$R_x(a, b, c, d) = \frac{a-c}{a-d} \cdot \frac{b-d}{b-c}$$

(In case one of a, b, c, d is ∞, one must make the definition more precise, e.g. if $a = \infty$, we get for the cross-ratio $\frac{b-d}{b-c}$.)

Theorem 8.5. Let F be a field, and let l, as above, be the projective line over F, with non-homogeneous coordinate x which varies over the set $F \cup \{\infty\}$. Then the group $PGL(l)$ of fractional linear transformations on F is precisely the group of permutations of l which preserve the cross-ratio, i.e. one-to-one mappings φ of l onto l, such that whenever A, B, C, D are four distinct points of l, and $\varphi(A) = A'$, etc. then

$$R_x(A, B, C, D) = R_x(A', B', C', D').$$

Proof: First we must see that every fractional linear transformation does preserve the cross-ratio. Since the group $PGL(l)$ is generated by transformations of the three special types (i), (ii), (iii) of Proposition 8.2, it will be sufficient to see that each one of them preserves the cross-ratio. So let A, B, C, D be four points of l, with coordinates a, b, c, d. Then

$$R_x(A, B, C, D) = \frac{a - c}{a - d} \cdot \frac{b - d}{b - c}.$$

(i) If we apply a transformation of the type $x' = x + \lambda$, $\lambda \in F$, our new points A', B', C', D' have coordinates $a + \lambda$, $b + \lambda$, $c + \lambda$, $d + \lambda$, respectively. Hence

$$R_x(A', B', C', D') = \frac{(a+\lambda) - (c+\lambda)}{(a+\lambda) - (d+\lambda)} \cdot \frac{(b+\lambda) - (d+\lambda)}{(b+\lambda) - (c+\lambda)},$$

which is easily seen to be equal to the original cross-ratio.

(ii) If we apply a transformation of the form $x' = \lambda x$, $\lambda \in F$, $\lambda \neq o$, we have

$$R_x(A', B', C', D') = \frac{\lambda a - \lambda c}{\lambda a - \lambda d} \cdot \frac{\lambda b - \lambda d}{\lambda b - \lambda c}$$

which again is clearly equal to the first cross-ratio.

(iii) If we apply the transformation $x' = \frac{1}{x}$, we have

$$R_x(A', B', C', D') = \frac{\frac{1}{a} - \frac{1}{c}}{\frac{1}{a} - \frac{1}{d}} \cdot \frac{\frac{1}{b} - \frac{1}{d}}{\frac{1}{b} - \frac{1}{c}}.$$

Now multiplying above and below by abcd, we obtain the original cross-ratio again. (One must consider the special case when one of a, b, c, d is o or ∞ separately - left to the reader).

Thus we have shown that every fractional linear transformation preserves the cross-ratio. Now conversely, let us suppose that φ is a transformation which preserves cross-ratio. Let φ send $o, 1, \infty$ into a, b, c respectively, and let $\varphi(x) = x'$. Then we have

$$R_x(o, 1, \infty, x) = R_x(a, b, c, x')$$

or

$$\frac{o - \infty}{o - x} \cdot \frac{1 - x}{1 - \infty} = \frac{a - c}{a - x'} \cdot \frac{b - x'}{b - c}$$

or

$$\frac{x - 1}{x} = \frac{a - c}{b - c} \cdot \frac{b - x'}{a - x'}$$

Solwing for x' , we find that φ is given by the expression

$$x' = \frac{\frac{a - b}{b - c} cx + a}{\frac{a - b}{b - c} x + 1} ,$$

which is indeed a fractional linear transformation.

q. e. d.

Example : Let $F = \mathbb{C}$ be the field of complex numbers. Then the line 1 , is the projective line over $\mathbb{C}$, that is, the "plane" of complex numbers , plus one additional point, called ∞. This is most easily represented as a sphere, called the Riemann sphere, via the stereographic projection. (For details, see any book on functions of a complex variable.) A unit sphere is placed on the origin of the complex plane (which becomes the S pole of the sphere) . Then projecting from the N pole of the sphere , the point at infinity corresponds to the N pole and all other points of the sphere correspond in a one-to-one manner with the points of the complex plane.

Now it is proved in courses on functions of a complex variable (q. v.) that the fractional linear transformations of the extended complex plane correspond precisely to those one-to-one transformations of the Riemann sphere onto itself which preserve orientation, and which are conformal, i. e. which preserve the angles between any two intersecting curves.

Projective Collineations.

Now we come to the study of projective collineations. In general, any automorphism of a projective plane π is called a collineation, because it sends lines into lines.

Definition : A projective collineation is an automorphism φ of the projective plane π, such that, whenever l is a line of π, and $l' = \varphi(l)$ is its image under φ, then the restriction of φ to l,

$$\varphi|_l : l \longrightarrow l' ,$$

which is a mapping of the line l to the line l', should be a projectivity.

For example, the identity transformation is a projective collineation. But we will see that in general, there are may more projective collineations. In fact we will prove that if π is a projective plane satisfying P 5 and P 6, then the projective collineations satisfy a fundamental theorem : there is a unique one of them sending any four points, no three collinear, into any other four points, no three collinear. We will also study the structure of the group of projective collineations, by showing that it is generated by certain special kinds of projective collineations, called elations and homologies. Finally, we will show that if $\pi \cong \mathbb{P}^2_F$, where F is a field, then the group of projective collineations is precisely $PGL(2, F)$.

Proposition 8.6 Let φ be an automorphism of π. Then φ is a projective collineation if and only if there exists some line l_o, such that $\varphi|_{l_o}$ is a projectivity.

Proof : If φ is a projective collineation, any l_o will do. So suppose conversely that φ is an automorphism whose restriction to l_o is a projectivity. Say $\varphi(l_o) = l_o'$. Now let l be any other line, and let P be a point not on l or l_o.

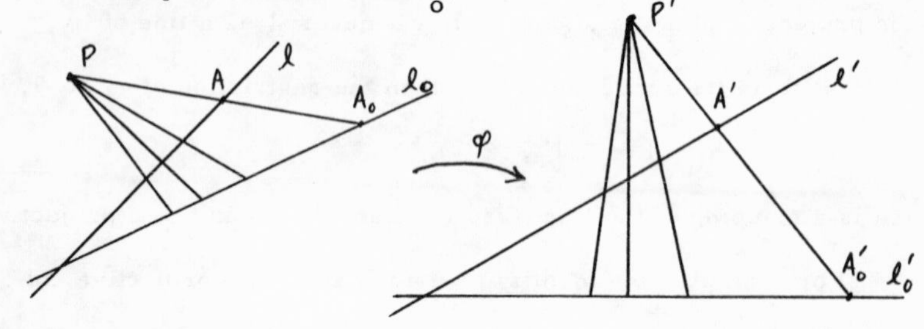

Let $\psi : l \longrightarrow l_o$ be the perspectivity $l \overset{P}{\underset{\wedge}{=}} l_o$. Now if $A \in l$ and $A_o \in l_o$, then say that $\psi(A) = A_o$ is the same as saying P, A, A_o are collinear. Since φ is an automorphism, this is the same as saying that P', A', A_o' are collinear (where ' denotes the action of φ). Let $l' = \varphi(l)$. In other words, the transformation

$$\varphi\psi\varphi^{-1} : l' \longrightarrow l_o'$$

is the same as the perspectivity $l' \overset{P'}{\underset{\wedge}{=}} l_o'$. Call it ψ'. So

$$\psi' = \varphi\psi\varphi^{-1} .$$

In other words,

$$\varphi|_l = \psi'^{-1}\varphi|_{l_o}\psi .$$

But $\psi, \varphi|_{l_o}$, and ψ'^{-1} are all projectivities, so $\varphi|_l$ is also a projectivity, and hence φ is a projective collineation, since l was arbitrary.

q.e.d.

Be fore we can prove much about projective collineations, we must study some special types of collineations, called elations and homologies. Then we will use them to deduce properties of the group of projective collineations.

Definition. An elation is an automorphism of the projective plane π, which leaves some line, say l_o, pointwise fixed, and which has no other fixed points. The line l_o is called the axis of the elation.

Let α be an elation of π, with axis l_o, and let A be the affine plane $\pi - l_o$. For any $P, Q \in A$, let PQ meet l_o at X. Then X is fixed, so $P'Q'$ also meets l_o at X, where P' and Q' are the images of P and Q under α. Hence $PQ \parallel P'Q'$ in A, so α restricted to A is a dilatation. But α has no fixed points outside of l_o, so α restricted to A is in fact a translation. Conversely any translation of A gives an elation of π with axis l_o.

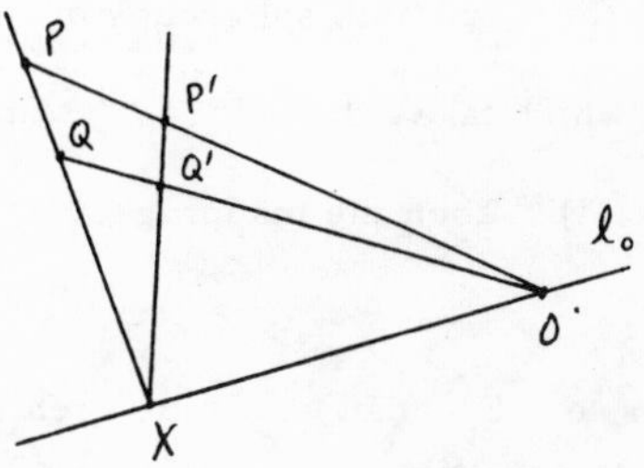

<u>Proposition 8.7</u> The elations of π with axis l_o correspond, by restriction, to the translations of the affine plane $\pi - l_o$. Hence, if one includes the identity, the elations with axis l_o form a group E_{l_o}.

<u>Proof :</u> We need only refer to the fact that the translations of an affine plane form a group.

If α is an elation with axis l_o, then we can speak of the direction of the translation $\alpha|_A$. Indeed, for any P, Q, $PP' \| QQ'$. Say they meet l_o at O. Then O is the <u>center</u> of the elation α.

One should not suppose that all the elations taken together form a group. For if α_1, α_2 are elations with different axes l_1, and l_2, there is no reason why $\alpha_1\alpha_2$ should be an elation at all.

However, we can say something about all the elations. First we have shown that the elations with a fixed axis l_o (including the identity) form a group, E_{l_o}. Similarly, if l_1 is another line, the elations are both subgroups of Aut π. Let φ be an automorphism of π which takes l_o into l_1 (so long as π satisfies P 5, there will be one !). Then the mapping

$$\alpha \longrightarrow \varphi\alpha\varphi^{-1}$$

for $\alpha \in E_{l_o}$ can easily be seen to be an isomorphism of E_{l_o} onto E_{l_1}. Note for example that φ^{-1} takes l_1 into l_o, α leaves l_o pointwise fixed, and φ takes l_o into l_1, so that $\varphi\alpha\varphi^{-1}$ leaves l_1 pointwise fixed. Similarly one can see that $\varphi\alpha\varphi^{-1}$ has no other fixed points, so it is an elation. We leave some details to the reader. This is a familiar situation in group theory. In fact, we have the following definition.

Definition . Let G be a group, and let H_o and H_1 be subgroups of G. Then we say that H_o and H_1 are conjugate subgroups, if there is an element $g \in G$, so that the map

$$h_o \longmapsto gh_og^{-1}$$

is an isomorphism of H_o onto H_1.

Thus we have proved

Proposition 8.8 Let π be a projective plane satisfying P 5. Let E_{l_o} and E_{l_1} denote the groups of elations of π with axes l_o and l_1, respectively. Then E_{l_o} and E_{l_1} are conjugate subgroups of Aut π.

Conversely, one can see easily that any conjugate subgroup of E_{l_o} is of the form E_l, for some line l in π. Thus the set of all elations of π is the union of the subgroup E_{l_o} of Aut π, together with its conjugates.

Definition . A homology of the projective plane π is an automorphism of π which leaves a certain line l_o pointwise fixed, and which has precisely one other fixed point O. l_o is called the axis of the homology, and O is called its center.

As above, we note that the homologies with axis l_o correspond to dilatations of the affine plane $\pi - l_o$. Hence, if one adjoins the elations with axis l_o, and the identity, they form a group, which we will call H_{l_o}. For any other axis l_1, H_{l_1} is a conjugate subgroup of Aut π to H_{l_o}.

Refining some more, we see that for any line l_o, and for any point O not on l_o, the homologies with axis l_o and center O form a group $H_{l_o,O}$. And since in a Desarguesian projective plane, we can move a line l_o and a point O to any other line l_1 and point P, we see as above that $H_{l_1,P}$ is conjugate to $H_{l_o,O}$. Hence the homologies of π are the union of the subgroup $H_{l_o,O}$ of Aut π with all of its conjugates.

Proposition 8.9. Elations and homologies are projective collineations.

Proof. By Proposition 8.6, it is sufficient to note that their restriction to a single line is a projectivity. But the restriction of any elation or homology to its axis is the identity, which is a projectivity.

Proposition 8.10. Let π be a projective plane satisfying P 5. Let A, B, C, D and A', B', C', D' be two quadruples of points, no three of which are collinear. Then one can find a product φ of elations and homologies, such that $\varphi(A) = A'$, $\varphi(B) = B'$, $\varphi(C) = C'$ and $\varphi(D) = D'$.

Proof: Step 1 Choose a line l_o such that A and A' are not on l_o. Then, since π is Desarguesian (cf. Chapter VII) there is a translation of $\pi - l_o$ which sends A into A', i.e. an elation α_1 of π such that $\alpha_1(A) = A'$. Let α_1 take B, C, D into B'', C'', D''. Then we have reduced to the problem of finding a product of elations and homologies which leaves A' fixed, and sends B'', C'', D'' into B', C', D'. Furthermore, since α_1 is an automorphism, A', B'', C'', D'' are four points no three of which are collinear. Thus, relabeling A', B'', C'', D'', as A, B, C, D , we have reduced to the original problem , under the additional assumption that A = A' .

Step 2. Choose another line l_1 such that $A \in l_1$, but $B, B' \notin l_1$. Then choose an elation α_2 with axis l_1 , and such that $\alpha_2(B) = B'$. Then using α_2, and relabeling again, we have reduced the original problem to the case A = A' and B = B'.

Step 3. Let l_2 = AB . Then C and C' are not on l_2 , because A, B, C are not collinear, and A', B', C' are not collinear. So again, we can choose an elation α_3 with axis l_2 , such that $\alpha_3(C) = C'$, and so reduce the problem to the case A = A', B = B', C = C'.

Step 4. Draw AD and BD' and let them meet at E. Now since A, D, E are collinear, and D, E, are different from A, There exists a dilatation of the affine plane π - BC , whcih leaves A fixed, and sends D into E. In other words, there is a homology β_1 of π with axis BC and center A, which sends D into E.

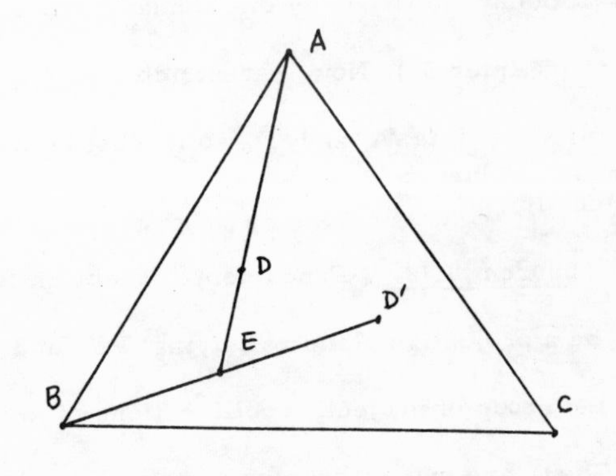

Step 5. Similarly, there is a homology β_2 of π with axis AC and center B, which sends E into D' . Therefore $\beta_2\beta_1$ leaves A, B, C

fixed, and sends D into D'.

This completes the proof of the proposition. Note that in general, we need three elations, and two homologies.

Proposition 8.11 Let π be a projective plane satisfying P 5 and P 6. Let φ be a projective collineation of π, which leaves fixed four points A, B, C, D, no three of which are collinear. Then φ is the identity.

Proof : Let l be the line BC. Since B and C are fixed, φ sends l into itslef, and φ restricted to l must be aprojectivity, since φ is a projective collineation. But φ also leaves A and D fixed, so φ must leave $AD.l = F$ fixed. So $\varphi|_l$ is a projectivity of l into itself which leaves fixed the three points B, C, F. Hence φ leaves l pointwise fixed, by the Fundamental Theorem for projectivities on a line (Chapter 5.). Now φ restricted to $\pi - l$ is a dilatation with two fixed points A and D, so it must be the identity. Hence φ is the identity.

Theorem 8.12. (Fundamental Theorem for Projective Collineations). Let π be a projective plane satisfying P 5 and P 6, and denote by $PC(\pi)$ the group of projective collineations of π. If A, B, C, D and A', B', C', D' are two quadruples of points, no three collinear, then there is a unique element $\varphi \in PC(\pi)$ such that $\varphi(A) = A'$, $\varphi(B) = B'$, $\varphi(C) = C'$, and $\varphi(D) = D'$.

Proof : Since elations and homologies are projective collineations

(Proposition 8.9) and since there are enough of them to send A, B, C, D to A', B', C', D' (Proposition 8.10), there certainly is some such φ. On the other hand, if ψ is another such projective collineation, then $\psi^{-1}\varphi$ is a projective collineation which leaves A, B, C, D fixed, and so is the identiy (Proposition 8.11). Hence $\varphi = \psi$, and φ is unique.

Corollary 8.13. The group $PC(\pi)$ of projective collineations is generated by elations and homologies.

Proof : Let $\psi \in PC(\pi)$, let A, B, C, D be four points, no three collinear, and let ψ send A, B, C, D into A', B', C', D'. Construct by Proposition 8.10 a product φ of elations and homologies which also sends A, B, C, D to A', B', C', D'. Then by the uniqueness of the theorem, $\psi = \varphi$, so ψ is a product of elations and homologies.

Finally, we come to the analytic interpretation of the projective collineations.

Theorem 8.14. Let F be a field, and let $\pi = \mathbb{P}^2_F$ be the projective plane over F. Then

$$PC(\pi) = PGL(2, F) .$$

Proof : First we will show that certain very special elations and homologies are represented by matrices.

Consider an elation α with axis $x_3 = o$ and center $(1, o, o)$. If A is the affine plane $x_3 \neq o$, with affine coordinates

$$x = x_1/x_3$$

$$y = x_2/x_3$$

then α is a translation of A in the x-direction, i.e. it has equations

$$x' = x + a \qquad\qquad a \in F.$$

$$y' = y$$

So its homogeneous equations are

$$x'_1 = x_1 + ax_3$$

$$x'_2 = x_2$$

$$x'_3 = x_3 ,$$

so α is represented by the matrix

$$E_a = \begin{pmatrix} 1 & o & a \\ o & 1 & o \\ o & o & 1 \end{pmatrix}$$

with $a \in F$.

Now if α' is any other elation, with axis l_o, and center O, we can find a matrix A, such that T_A sends the line $x_3 = o$ into l_o, and (1, o, o) to O. Then α' will be of the form

$$\alpha' = T_A \, \alpha \, T_A^{-1} ,$$

where α is an elation of the above special type. In other words, α' is represented by the matrix AE_aA^{-1}, for some $a \in F$.

Similarly, consider a homology β, with axis $x_1 = o$ and center (1, o, o). Passing to the affine plane $x_1 \neq o$, we see that it is a dilatation with center (o, o), hence is a streching in some ratio $k \neq o$, and its equation in homogeneous coordinates is

$$x'_1 = x_1$$

$$x'_2 = kx_2$$

$$x'_3 = kx_3 \ .$$

So it is represented by the matrix

$$\begin{pmatrix} 1 & o & o \\ o & k & o \\ o & o & k \end{pmatrix} .$$

We can get another matrix representing the same transformation by multiplying by the scalar $b = k^{-1}$, so we find β is represented also by the matrix

$$H_b = \begin{pmatrix} b & o & o \\ o & 1 & o \\ o & o & 1 \end{pmatrix} \qquad b \in F, \ b \neq o.$$

As before, any other homology β' is a conjugate by some matrix B of one of this form, so any homology β' is represented by a matrix of the form BH_bB^{-1} , for some $b \in F$, $b \neq o$.

Thus we have seen that every elation and every homology can be represented by a matrix, i.e. they are elements of the group PGL(2, F) . But by Corollary 8.13 above, the group of projective collineations is generated by elations and homologies, so we have

$$PC(\pi) \subseteq PGL(2, F) .$$

But we have seen (Chapter VI) , that over a field F, there is a unique element of PGL(2, F) sending four points, no three collinear, into four points, no three collinear. Since this is already accomplished by the subgroup $PC(\pi)$, according to the Fundamental Theorem above, the two groups must be equal.

q.e.d.

Corollary 8.15 . Let F be a field. Then every invertible 3×3 matrix M with coefficients in F can be written as a scalar times a product of conjugates of matrices of the two forms E_a and H_b above. In particular, we can write M in the form

$$M = \lambda B_2 H_{b_2} B_2^{-1} B_1 H_{b_1} B_1^{-1} A_3 E_{a_3} A_3^{-1} E_{a_2} A_2^{-1} A_1 E_{a_1} A_1^{-1}$$

with $a_1, a_2, a_3 \in F$, $b_1, b_2, \lambda \in F$, $b_1, b_2, \lambda \neq 0$, A_1, A_2, A_3, B_1, B_2 invertible matrices.

Remark : From this result, one can deduce with comparatively little effort the fact that the determinant function on 3×3 matrices is determined uniquely by the properties D 1 and D 2 on page 30 . Compare also Problem 19.

PROBLEMS

In the following problems, you may use the axioms and propositions given in class. Refer to them explicitly.

1. Show that any two pencils of parallel lines in an affine plane have the same cardinality (i. e. that one can establish a one-to-one correspondence between them). Show that this is also the cardinality of the set of points on any line.

2. If there is a line with exactly n points, show that the number of points in the whole affine plane is n^2.

3. Discuss the possible systems of points and lines which satisfy P 1, P 2, P 3, but not P 4.

4. Prove that the projective plane of 7 points, obtained by completing the affine plane of four points, is the smallest possible projective plane..

5. If one line in a projective plane has n points, find the number of points in the projective plane.

6. Let S be a projective plane, and let l be a line of S. Define S_o to be the points of S not on l, and define lines in S_o to be the restrictions of lines in S. Prove (using P 1 - P 4) that S_o is an affine plane. Prove also that S is isomorphic to the completion of the affine plane S_o.

7. Using the axioms S 1 - S 6 of projective three-space, prove the following statements. Be very careful not to assume anything except what is stated by the axioms. Refer to the axioms explicitly by number.

a) If two distinct points P, Q lie in a plane Σ then the line joining them is contained in Σ.

b) A plane and a line not contained in the plane meet in exactly one point.

c) Two distinct planes meet in exactly one line.

d) A line and a point not on it lie in a unique plane.

8. Prove that any plane Σ in a projective three-space is a projective plane, i.e. satisfies the axioms P 1 - P 4. (You may use the results of the previous problem.)

FINITE AFFINE PLANES

9. Show that any two affine planes with 9 points are isomorphic. (We say that two planes A and A' are isomorphic if there is a one-to-one mapping $T : A \longrightarrow A'$, which takes lines into lines.)

10. Construct an affine plane with 16 points. (Hint: We know from problem 1 that each pencil of parallel lines has four lines in it. Let a, b, c, d be one pencil of parallel lines, and let 1, 2, 3, 4 be another. Then label the intersections $A_1 = a \cap 1$, etc. To construct the plane, you must choose other subsets of four points to be the lines in the three other pencils of parallel lines. Write out each line explicitly, by naming its four points, e.g. the line $2 = \{ A_2, B_2, C_2, D_2 \}$.

11. Euler in 1779 posed the following problem :

" A meeting of 36 officers of six different ranks and from six different regiments must be arranged in a square in such a manner that each row and each column contains 6 officers from different regiments and of different ranks " .

It has been shown that this problem has no solution. Deduce from this fact that there is no affine plane with 36 points.

We will consider the Desargues configuration, which is a set of 10 elements, $\Sigma = \{ O, A, B, C, A', B', C', P, Q, R \}$, and 10 lines, which are the subsets

O, A, A'
O, B, B'
O, C, C'
A, B, P
A', B', P
A, C, Q
A', C', Q
B, C, R
B', C', R
$P, Q, R.$

Let $G = \text{Aut } C$ be the group of automorphisms of Σ .

12. Show that G is transitive on Σ.

13. a) Show that the subgroup of G leaving a point fixed is transitive on a set of six letters.

b) Show that the subgroup of G leaving two collinear points fixed has order 2.

c) Deduce the order of G from the previous results.

Now we consider some further subsets of Σ, which we call planes, namely

$$1 = \{ O, A, B, A', B', P \}$$
$$2 = \{ O, A, C, A', C', Q \}$$
$$3 = \{ O, B, C, B', C', R \}$$
$$4 = \{ A, B, C, P, Q, R \}$$
$$5 = \{ A', B', C', P, Q, R \}$$

14. Show that each element of G induces a permutation of the set of five planes, $\{1, 2, 3, 4, 5\}$, and that the resulting mapping

$$\varphi : G \longrightarrow \text{Perm}\{1, 2, 3, 4, 5\}$$

is an isomorphism of groups. Thus G is isomorphic to the permutation group on five letters.

15. a) Let π_o be a set of four points A, B, C, D, and no lines. Let π be the free projective plane generated by the configuration π. (as in class). Show that any permutation of the set $\{A, B, C, D\}$ extends to an automorphism of the projective plane π.

b) Show that these are not the only automorphisms of π.

16. Prove that there is no finite configuration in the real projective plane such that each line contains at least three points, every pair of distinct points lies on a line, and not all the points are collinear. (Hint: First reduce to the Euclidean plane, then choose a triangle with minimal altitude.)

17. Let π be a projective plane. Let T be an involution of π, that is, let T be an automorphism of π such that $T^2 = T \cdot T =$ identity map of π. Let Σ be the set of fixed point of π. Prove that one (and only one) of the following is true :

Case 1. There is a line l_o in π such that $\Sigma = l_o$.

Case 2. There is a line l_o and a point $P_o \notin l_o$, such that $\Sigma = l_o \cup \{P_o\}$.

Case 3. Σ is a projective plane, where we define a "line" in Σ to be any subset of Σ, of the form (line in π) $\cap\, \Sigma$, which has at least two points.

Prove furthermore that Case 1. can arise only if the axiom P 7 is not satisfied.

18. For each case 1, 2, 3 above, give without proof a specific example of a projective plane π, and an involution $T \neq$ identity, which has the property of the given case.

19. Let φ be a function from the set of 2×2 real matrices $\{ A = \begin{pmatrix} a & b \\ c & d \end{pmatrix} \}$ to the real numbers, such that

D 1. $\varphi(A \cdot B) = \varphi(A) \cdot \varphi(B)$, and

D 2. $\varphi\begin{pmatrix} a & o \\ o & 1 \end{pmatrix} = a$, for each $a \in \mathbb{R}$.

Prove that $\varphi(A) = \det A$, i.e. $\varphi\begin{pmatrix} a & b \\ c & d \end{pmatrix} = ad - bc$, for all $a, b, c, d \in \mathbb{R}$.

(A similar but more involved proof would work for $n \times n$ matrices).

20. Let π be the real projective plane, and let

$$A = (a, o, 1)$$

$B = (b, o, 1)$

$C = (c, o, 1)$

$D = (d, o, 1)$, $a, b, c, d \in \mathbb{R}$,

be four points on the " x_1-axis" . Prove that AB, CD are four harmonic points if and only if the product

$$R_x(AB, CD) \equiv \frac{a-c}{a-d} \cdot \frac{b-d}{b-c}$$

is equal to -1 . (In general, this product $R_x(AB, CD)$ is called the <u>cross-ratio</u> of the four points.) You may use methods of Euclidean geometry in the affine plane $x_3 \neq o$.

21. By interchanging the words "point" and "line" , etc., make a careful statement of the dual, P 6* , of Pappus' axiom, P 6. Then use P 1 - P 4 and P 6 to prove P 6* .

22. Consider the configuration of Pappus' axiom in the real projective plane, and take the line PQ (using the notation given in class) to be the line at infinity. Pappus' axiom then becomes a statement in the Euclidean plane. Write out this statement, and then prove it, using methods of Euclidean geometry. (This gives a second proof that P 6 holds in the real projective plane) .

For the next three problems, we consider the following situation : let

$$l \overset{O}{\underset{\wedge}{=}} m \overset{P}{\underset{\wedge}{=}} n$$

be a chain of two perspectivities , and assume $l \neq n$. Let $\varphi : l \longrightarrow n$ be the resulting projectivity from l to n, and let X be the point $l \cdot n$.

23. 1) Prove that if φ is actually a perspectivity , then $\varphi(X) = X$.

2) Now assume simply that $\varphi(X) = X$, and prove that one of the following conditions holds :

a) l, m, n are concurrent, or

b) O, P, X are collinear.

24. With the initial hypotheses above , assume furthermore that l, m, n are concurrent. Prove that there is a point Q such that O, P, Q are collinear, and φ is the perspectivity $l \overset{Q}{\overline{\overline{\wedge}}} n$. (Use P5 or P5*).

25. With the initial hypotheses above, assume also that O, P, X are collinear, but that l, m, n are not concurrent. Let $Y = l \cdot m$, let $Z = m \cdot n$, and let $Q = OZ \cdot PY$. Prove that φ is the perspectivity $l \overset{Q}{\overline{\overline{\wedge}}} n$. (Use P 6 or P 6*) .

Remark : The problems 23, 24, 25 give a proof of lemma 5.4 mentioned in class. In fact, they prove a stonger result, namely, that under the initial hypotheses above, the following three conditions are equivalent :

(i) φ is a perspectivity

(ii) $\varphi(X) = X$

(iii) either a) or b) of # 23 above is true.

26. Let $k = \{o, 1, 2\}$ be the field of 3 elements, with addition and multiplication modulo 3. Let $F = \{a + bj \mid a, b \in k\}$, and where j is a new symbol.

a) Define addition and multiplication in F, using the relation $j^2 = 2$, and prove that F is then a field.

b) Prove that the multiplicative group F^* of non-zero elements of F is cyclic of order 8.

27. Let $A = F$ as a set, and denote the elements of A as (x) where $x \in F$. Define addition and multiplication in A as follows :

$(x) + (y) = (x + y)$ (here the left-hand + is the addition in A; the right-hand + is the addition in F).

$$(x)(y) = \begin{cases} (xy) & \text{if } y \text{ is a square in } F \\ (x^3 y) & \text{if } y \text{ is not a square in } F. \end{cases}$$

(We say y is a square in F if $\exists\, z \in F$ such that $y = z^2$).

Prove a) A is an abelian group under +

b) The non-zero elements A^* of A from a group under multiplication.

c) $(o)(x) = (x)(o) = (o)$ for all $(x) \in A$

d) $((x) + (y))(z) = (x)(z) + (y)(z)$ for all (x), $(y), (z) \in A$.

28. Let A be a finite algebra satisfying a), b), c), d) of the previous problem (i.e. A is a finite set, with two operations t, such that a), b), c), d) hold.). Note that A would be a division ring, except that the left distributive law is missing. Prove that one can construct

a projective plane $\mathbb{P}_A^2$ over A as follows :

I. A point is an equivalence class of triples (x_1, x_2, x_3) with $x_i \in A$, where $(x_1, x_2, x_3) \sim (x_1\lambda, x_2\lambda, x_3\lambda)$ for any $\lambda \in A$, $\lambda \neq o$. (Prove this is an equivalence condition)

II . A line is the set of all points satisfying an equation of the form

$$x_1 + bx_2 + cx_3 = o \qquad b, c \in A$$

or

$$x_2 + cx_3 = o \qquad c \in A$$

or

$$x_3 = o \, .$$

(Prove these equations determine sets of points)

III Verify the axioms P 1 - P 4 .

Two warnings : 1) Not all linear equations define lines !

2) You may need to use the finiteness of A somewhere in the proof.

29. If A is the algebra of the problem 27, show that $\mathbb{P}_A^2$ does not satisfy Desargues axiom P 5. Thus $\mathbb{P}_A^2$ is an example of a finite non-Desarguesean projective plane.

30. Axioms for the real affine plane

In the ordinary Euclidean plane, let <ABC> stand for the relation " A, B, C are collinear, and B is between A and C ". Write down some nice properties of this relation.

Now let Σ be an abstract affine plane satisfying A1, A2, A3, A5a, A5b, and A6 (you define this one-Pappus' axiom). Assume that

Σ has a notion of betweenness given, i.e. for certain triples of points $A, B, C \in \Sigma$, we have $< ABC >$, and assume that this notion $< \ >$ satisfies certain axioms, namely the properties you listed earlier. (Make sure there were enough). Add further a "completeness" axiom, say

C : whenever a line l is devided into two non-empty subsets l' and l'', so that no element of one subset is between two elements of the other subset, then there exists a unique point $A \in l$, such that $\forall B \in l'$, $\forall C \in l''$, $B \neq A$ and $C \neq A$, we have $< BAC >$. (Dedekind cut axiom).

Now try to prove that your geometry Σ, with this notion of betweenness, must be the affine plane over the real numbers $\mathbb{R}$. (You may use the theorem that $\mathbb{R}$ is the only complete ordered field.)

Hint: Try the following as one of your axioms (Pasch's axiom) : If A, B, C are three non-collinear points, and if

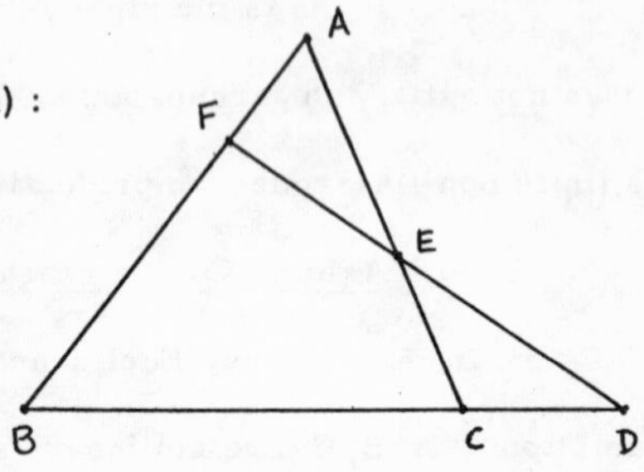

$$< BCD >$$

and

$$< AEC >,$$

then there exists a point F on the line DE, such that $< BFA >$.

31. Let S_4 be the group of permutations of the four symbols 1, 2, 3, 4.

a) Let $G \subseteq S_4$ be the subgroup generated by the permutation (1234). What is the order of G? (The order is the number of elements in G).

b) Let $H \subseteq S_4$ be the subgroup generated by the permutations (12) and (34). What is the order of H ?

c) Is there an isomorphism (of abstract groups) $\varphi : G \longrightarrow H$? If so, write it explicitly. If not, explain why not.

32. The Pappus Configuration. Σ, is the configuration of 9 points and 9 lines as hown in the diagram

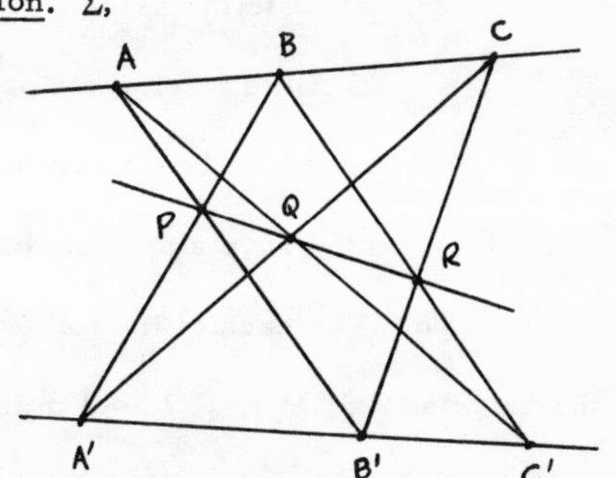

a) What is the order of the group of automorphism of Σ ?

b) Explain briefly how you arrived at the answer to a).

33. a) In the real projective plane, what is the equation of the line joining the points (1, 0, 1) and (1, 2, 3) ?

b) What is the point of intersection of the lines

$$x_1 - x_2 + 2x_3 = 0$$

$$3x_1 + x_2 + x_3 = 0 \quad ?$$

34. In the real projective plane, we know that there is an automorphism which will send any four points, no three collinear, into any four points, no three collinear. Find the coefficients a_{ij} of an automorphism with equations

$$x'_i = \sum_{j=1}^{3} a_{ij} x_j \qquad i = 1, 2, 3$$

which sends the points

$$A = (0,0,1)\ ,\quad B = (0,1,0)\ ,\quad C = (1,0,0)\ ,\quad D = (1,1,1)$$

into

$$A' = (1,0,0)\ ,\ B' = (0,1,1)\ ,\quad C' = (0,0,1)\ ,\ D' = (1,2,3)$$

respectively.

35. a) State the axioms P 1, P 2, P 3, P 4 of a projective plane.

b) Give a complete proof that they imply the statement

Q : "There are four points, no three of which are collinear".

c) Prove also that P 1, P 2, and Q imply P 3 and P 4.

36. For each of the following projective planes, state which of the axioms P 5, P 6, P 7 hold in it, and explain why each axiom does or does not hold. (Please refer to results proved in class, and give brief outlines of their proofs.)

a) The projective plane of seven points.

b) The real projective plane.

c) The free projective plane generated by four points.

37. a) Draw a picture of the projective plane of seven points, π.

b) I there an automorphism T of π such that T^7 = identity, but $T \neq$ identity ? If so, write one down explicitly. If not, explain why not.

38. Let l, l' be two distinct lines in a projective plane π. Let $X = l \cdot l'$. Let A, B be two distinct points on l, different from X. Let C, D be two distinct points on l', different from X. Construct a projectivity $\varphi : l \longrightarrow l'$ which sends A, X, B into X, C, D, respectively.

39. Let l be a line in a projective plane π satisfying P 1 - P 6. Let φ be a permutation of the points on l, such that for any four points A, B, C, D on l, AB, CD are four harmonic points $\Leftrightarrow$ $A'B'$, $C'D'$ are four harmonic points (where $A' = \varphi(A)$, $B' = \varphi(B)$ etc.). Is φ necessarily a projectivity of l into itself ? Proof or counter example.

40. Find the diagonal points of the complete quadrangle on the four points $(\pm 1, \pm 1, 1)$.

41. Let π be a projective plane of seven points. Let A and B be two distinct points of π. How many automorphisms of π are there which send A to B ? Give your reasons !.

42. a) Let F be a division ring, and let λ be a fixed non-zero element of F. Prove that the map $\varphi : F \longrightarrow F$, defined by

$$\varphi(x) = \lambda x \lambda^{-1}$$

for all $x \in F$, is an automorphism of F.

b) Let p be a prime number. Prove that the field F of p elements has no automorphisms other than the identity automorphism. (Recall that $F = \{0, 1, \ldots, p-1\}$, where addition and multiplication are defined modulo p.)

43. Let F be the field with three elements, let $\pi = \mathbb{P}_F^{\ 2}$, and let ℓ be any line of π. Show that ℓ has exactly four points A,B,C,D and that they are four harmonic points, in any order. Quote explicitly any theorems from class which you may wish to use.

44. In the ordinary Euclidean plane (considered as being contained in the real projective plane), let C be a circle with center O, let P be a point outside C, and let t_1 and t_2 be the tangents from P to C, meeting C at A_1 and A_2. Draw A_1A_2 to meet OP at B, and let OP meet C at X and Y. Prove (by any method) that X,Y,B,P are four harmonic points.

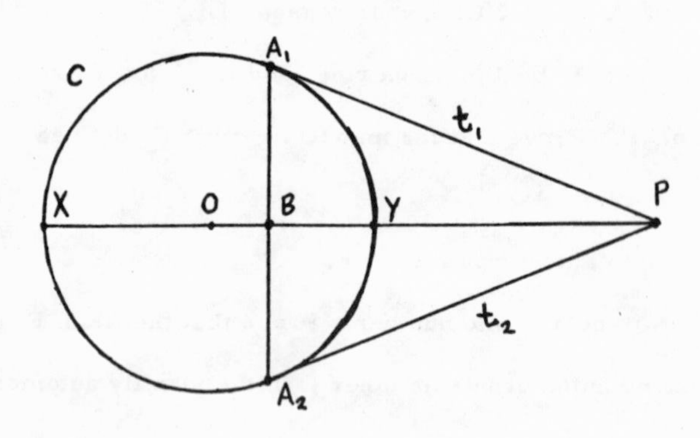

45. Let F be a field, and let $X = (x_1,x_2,x_3)$, $Y = (y_1,y_2,y_3)$ and $Z = (z_1,z_2,z_3)$ be three points in the projective plane $\pi = \mathbb{P}_F^{\ 2}$. If $X \neq Y$, and X,Y,Z are collinear, prove that there exist elements λ and μ in F such that

$$z_i = \lambda x_i + \mu y_i \qquad \text{for } i = 1,2,3.$$

46 . Let π be a projective plane satisfying P5, P6, and P7, and let ℓ be a line in π. Prove that if φ is a projectivity of ℓ into ℓ which interchanges two distinct points A,B of ℓ (i.e. $\varphi(A) = B$ and $\varphi(B) = A$) , then φ^2 is the identity.

<u>Hint</u>: Let C be another point of ℓ and let $\varphi(C) = D$. Construct a projectivity $\psi : \ell \to \ell$ which interchanges A and B , and interchanges C and D , using the diagram below. Then apply the Fundamental Theorem.

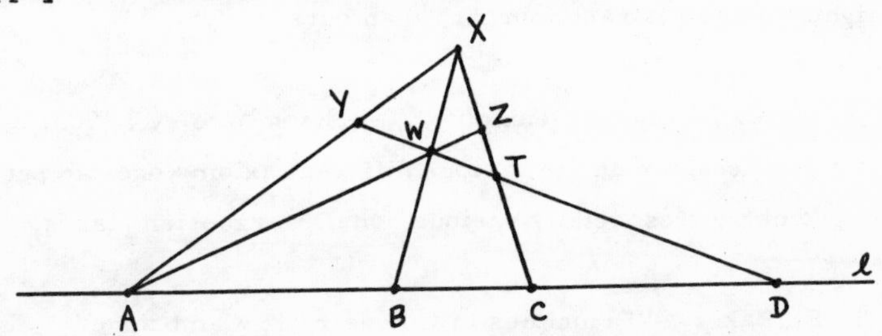

47. Let p be a prime number, let F be the field with p elements, let $\pi = \mathbb{P}^2_F$, and let $G = \text{Aut}\ \pi$. Prove that the order of G is $p^3(p^3-1)(p^2-1)$.

Hints: First prove that $G = PGL(2,F)$. Then use the result from class which says that a matrix

$$\begin{pmatrix} a_1 & a_2 & a_3 \\ b_1 & b_2 & b_3 \\ c_1 & c_2 & c_3 \end{pmatrix}$$

of elements of F has determinant $\neq 0$ if and only if no row is all zeros, and the points $A = (a_1,a_2,a_3)$, $B = (b_1,b_2,b_3)$, and $C = (c_1,c_2,c_3,)$ of π are not collinear. Or you may use the Fundamental Theorem for projective collineations of π .

BIBLIOGRAPHY

This list includes only works which were used directly in preparing the present notes. Refer to them for further bibliography .

1. E. Artin , "Geometric Algebra" , Interscience , N.Y. 1957.
- chapter II contains the construction of coordinates in an affine plane, from a slightly more abstract approach than ours.

2. R. Artzy , "Linear Geometry" , Addison-Wesley , 1965.
- contains a good chapter on the various different axioms one can put on a plane geometry , especially various non-Desarguesian planes.

3. H.F. Baker , "Principles of Geometry" , Cambridge University 1929-1940 .
- Volume I , chapter I . has the proof that any chain of perspectivities between distinct lines can be reduced to a chain of length two.

4. G. Birkhoff and S. MacLane , "A survey of Modern Algebra" Macmillan , 1941 .
- we refer to the chapter on group theory to supplement the very sketchy treatment given in these notes.

5. R.D. Carmichael , "Introduction to the theory of groups of finite order" , 1937 , Dover reprint,1956 .
- section 108 contains examples of finite non-Desarguesian projective planes, one of which we have reproduced in problems 26-29.

6. H.S.M. Coxeter , "The Real Projective Plane" , McGraw-Hill , 1949 .
- A good general reference for synthetic projective geometry.

7. H.S.M. Coxeter , "Introduction to Geometry" , Wiley , 1961.
- chapter 14, gives a good brief survey of the basic topics of projective geometry .

8. W.T. Fishback , "Projective and Euclidean Geometry" , Wiley , 1962 .
- A good general reference , much in the spirit of our treatment.

9. D. Hilbert and S. Cohn-Vossen , "Geometry and the Imagination". Chelsea,1952 (translated from German , "Anschauliche Geometrie" , Springer 1932.)
- chapter III on projective configurations is very pleasant reading and quite relevant.

10. M. Kraitchik , "Mathematical Recreations" , Norton Co., 1942. Dover reprint 1953 .
- see chapter VII , section 12 for the interpretation of magic squares as finite affine planes , and Euler's problem of the officers.

11. A. Seidenberg , "Lectures in Projective Geometry" , Van Nostrand , 1963.
- a very good general reference , with emphasis on axiomatics.

[illegible]

[illegible]

[illegible]

[illegible]

[illegible]

[illegible]

[illegible] Mathematical [illegible]

[illegible]

[illegible]